T0210710

Lecture Notes in Computer Science 9565

Commenced Publication in 1973
Founding and Former Series Editors:
Gerhard Goos, Juris Hartmanis, and Jan van Leeuwen

Editorial Board

More information about this series at http://www.springer.com/series/7410

Moti Yung · Jianbiao Zhang
Zhen Yang (Eds.)

Trusted Systems

7th International Conference, INTRUST 2015
Beijing, China, December 7–8, 2015
Revised Selected Papers

 Springer

Editors
Moti Yung
Computer Science
Columbia University
New York, NY
USA

Zhen Yang
Key Laboratory of Trusted Computing
Beijing University of Technology
Beijing
China

Jianbiao Zhang
Key Laboratory of Trusted Computing
Beijing University of Technology
Beijing
China

ISSN 0302-9743   ISSN 1611-3349 (electronic)
Lecture Notes in Computer Science
ISBN 978-3-319-31549-2   ISBN 978-3-319-31550-8 (eBook)
DOI 10.1007/978-3-319-31550-8

Library of Congress Control Number: 2016934019

LNCS Sublibrary: SL4 – Security and Cryptology

Printed on acid-free paper

This Springer imprint is published by Springer Nature
The registered company is Springer International Publishing AG Switzerland

Preface

These proceedings contains the 15 papers presented at the INTRUST 2015 conference, held in Beijing, China, in December 2015. INTRUST 2015 was the 7th International Conference on the Theory, Technologies and Applications of Trusted Systems. It was devoted to all aspects of trusted computing systems, including trusted modules, platforms, networks, services, and applications, from their fundamental features and functionalities to design principles, architecture, and implementation technologies. The goal of the conference was to bring academic and industrial researchers, designers, and implementers together with end-users of trusted systems, in order to foster the exchange of ideas in this challenging and fruitful area.

INTRUST 2015 built on the six previous successful INTRUST conferences: INTRUST 2009, INTRUST 2010, and INTRUST 2011 (Beijing, China), INTRUST 2012 (London, UK), INTRUST 2013 (Graz, Austria), and INTRUST 2014 (Beijing, China). The proceedings of all the previous INTRUST conferences have been published in the *Lecture Notes in Computer Science* series by the Springer.

Apart from the 15 contributed papers, the program of INTRUST 2015 also included four keynote speeches from Robert Deng (Singapore Management University), Wenchang Shi (Renmin University), Rob Spiger (Microsoft), and Claire Vishik (Trusted Computing Group). Special thanks are due to these keynote speakers.

The contributed papers were selected from 29 submissions from 11 countries. All submissions were blind-reviewed, i.e., the Program Committee members provided reviews on anonymous submissions. The refereeing process was rigorous, involving on average three (and mostly more) independent reports being prepared for each submission. The individual reviewing phase was followed by profound discussions about the papers, which contributed greatly to the quality of the final selection. We are very grateful to our hard-working and distinguished Program Committee for doing such an excellent job in a timely fashion.

For the proceedings, the papers have been arranged in four main categories, namely, encryption and signatures, security model, trusted technologies, and software and system security.

We also want to thank the conference Steering Committee, including Liqun Chen, Robert Deng, Yongfei Han, Chris Mitchell, and Moti Yung, the conference general chairs, Liqun Chen and Yongfei Han, the conference organizing chair, Lijuan Duan, the publicity chairs, Liqun Chen and Li Lin, and the Organizing Committee members, including Jing Zhan, Yingxu Lai, Zhen Yang, Yihua Zhou, Bei Gong, and Wei Ma, for their valuable guidance and assistance and for handling the arrangements in Beijing. Thanks are also due to the developers of EasyChair for providing the submission and review webserver.

On behalf of the conference organization and participants, we would like to express our appreciation to Beijing University of Technology and ONETS Wireless and Internet Security Company for their generous sponsorship of this event.

We would also like to thank all the authors who submitted their papers to the INTRUST 2015 conference, all external reviewers, and all the attendees of the conference. Authors of accepted papers are thanked again for revising their papers according to the feedback from the conference participants. The revised versions were not checked by the Program Committee, and thus authors bear full responsibility for the content. We thank the staff at Springer for their help with the production of the proceedings.

January 2016   Moti Yung
                                                          Jianbiao Zhang
                                                             Zhen Yang

INTRUST 2015

The 7th International Conference on Trusted Systems

Beijing, P.R. China

December 7–8, 2015

Sponsored by

Beijing University of Technology

ONETS Wireless & Internet Security Company

General Chairs

Liqun Chen   Hewlett Packard Laboratories, UK
Yongfei Han   BJUT and ONETS, China

Program Chairs

Moti Yung   Google and Columbia University, USA
Qiang Tang   University of Luxembourg, Luxembourg
Jianbiao Zhang   Beijing University of Technology, China

Program Committee

Liqun Chen   Hewlett Packard Laboratories, UK
Chris Mitchell   RHUL, UK
Sasa Radomirovic   ETH, Zurich, Switzerland
Jean Lancrenon   University of Luxembourg, Luxembourg
David Galindo   Scytl, Spain
Zhen Han   Beijing Jiaotong University, China
Wenchang Shi   Renmin University, China
Yanjiang Yang   Institute for Infocomm Research, Singapore
Liehuang Zhu   Beijing Institute of Technology
Jing Zhan   Beijing University of Technology, China
Endre Bangerter   Bern University of Applied Sciences, Switzerland
Zhen Chen   Tsinghua University, China
Zhong Chen   Peking University, China
Dieter Gollmann   Hamburg University of Technology, Germany
Sigrid Guergens   Fraunhofer Institute for Secure Information Technology,
                         Germany
Weili Han   Fudan University, China
Xuejia Lai   Shanghai Jiaotong University, China
Shujun Li   University of Surrey, UK
Jianxin Li   Beihang University, China
Peter Lipp   Graz University of Technology, Austria

Publicity Chairs

Liqun Chen   Hewlett Packard Laboratories, UK
Li Lin   Beijing University of Technology, China

External Reviewers

Bei Gong   Guan Wang
Jun Hu   Yuguang Yang
Wei Jiang   Xinlan Zhang
Yingxu Lai   Yihua Zhou
Weimin Shi

Contents

Trusted Technologies

Software and System Security

Encryptions and Signatures

Privacy-Preserving Anomaly Detection Across Multi-domain for Software Defined Networks

Huishan Bian[1], Liehuang Zhu[1], Meng Shen[1,2](✉), Mingzhong Wang[3], Chang Xu[1], and Qiongyu Zhang[1]

[1] Beijing Engineering Research Center of High Volume Language Information Processing and Cloud Computing Applications, School of Computer Science, Beijing Institute of Technology, Beijing, People's Republic of China
{heiseon,liehuangz,shenmeng,xuchang,biterzqy}@bit.edu.cn
[2] Ministry of Education, Key Laboratory of Computer Network and Information Integration (Southeast University), Nanjing, People's Republic of China
[3] Faculty of Arts and Business, University of the Sunshine Coast, Queensland, Australia
mwang@usc.edu.au

Abstract. Software Defined Network (SDN) separates control plane from data plane and provides programmability which adds rich function for anomaly detection. In this case, every organization can manage their own network and detect anomalous traffic data using SDN architecture. Moreover, detection of malicious traffic, such as DDoS attack, would be dealt with much higher accuracy if these organizations shared their data. Unfortunately, they are unwilling to do so due to privacy consideration. To address this contradiction, we propose an efficient and privacy-preserving collaborative anomaly detection scheme. We extend prior work on SDN-based anomaly detection method to guarantee accuracy and privacy at the same time. The implementation of our design on simulated data shows that it performs well for network-wide anomaly detection with little overhead.

Keywords: Privacy-preserving · Multi-domain collaboration · Anomaly detection · Software defined network

1 Introduction

Software Defined Network (SDN) separates the control plane from the data plane which provides rich functionality. In SDN environment, network operators can utilize this characteristic to manage their network domain. There is a logically centralized controller in every domain which consists of one or more physical controllers. The logically centralized controller is responsible for dealing with all intra-domain and inter-domain network events. Another essential function of controller is that it is responsible for anomaly detection in order to ensure SDN domain works well.

Several anomaly detection methods have been presented to protect security of domain in SDN environment [4,5,18,20], such as entropy based anomaly detection [4], SOM-based (Self Organizing Maps) [5] anomaly detection, etc.

© Springer International Publishing Switzerland 2016
M. Yung et al. (Eds.): INTRUST 2015, LNCS 9565, pp. 3–16, 2016.
DOI: 10.1007/978-3-319-31550-8_1

However, existing anomaly detection methods are constrained to single domain. Rather than considering the traffic of each domain independently, analyzing the traffic on all domains is more efficient because many abnormal behaviors impact multiple domains. Consequently, anomaly detection would be much easier if multiple domains shared their inner information, such as traffic traces, packets information, etc. This kind of information is vital for executing anomaly detection algorithms. It has been shown that if every domain cooperated with each other, one could detect anomaly that couldn't be detected by single domain [7]. Unfortunately, some domains are unwilling to disclose the details of intra-domains' information [1,8] since they are extremely confidential and information disclosure will cause direct attacks. Therefore, there should be a scheme that guarantees privacy without much performance sacrifice.

To address this problem, we propose an efficient and privacy-preserving collaborative anomaly detection scheme across multiple domains. We focus on the privacy issue, so we just utilize existing anomaly detection method to detect the typical attack, such as DDoS attack, in this paper. We adopt an effective anomaly detection method, which called SOM-based algorithm [5] to detect anomaly. On the basis, we perturb private information with random value and eliminate random value in subsequent computation which can protect privacy without sacrificing accuracy. In our scheme, all participants can learn the detection results without learning others' private information. We evaluate our design at a simulated distributed environment and the results show that our scheme performs well for network-wide anomaly detection.

The main contributions of this paper are as follows:

1. We present a problem of privacy disclosure risk which exists on multiple domains for SDN environment. To the best of our knowledge, it is the first attempt to consider privacy on multiple domains for SDN environment.
2. We propose a solution for dealing with the aforementioned problem. In our work, multiple domains can collaborate to detect network-wide malicious traffic without any leak of private information.
3. We evaluate our design, the results indicate that the design has a good performance for network-wide anomaly detection whether efficiency or accuracy.

The paper is structured as follows: In Sect. 2, we review the previous work and the main differences with our proposed mechanism, while in Sect. 3 we describe the system design. Section 4 presents our scheme in detail. Privacy analysis and performance evaluation are described in Sect. 5.

2   Related Work

Anomaly detection has been studied widely and has received considerable attention recently. Compared with legacy networks, there are many advantages on SDN infrastructure. For example, there is no need to deploy extra devices for anomaly detection on SDN and SDN infrastructure can provide fast mitigation. Therefore, many researchers study anomaly detection in SDN infrastructure to improve its flexibility and efficiency. In literature [5], the author proposed a lightweight method to extract traffic flow features for DDoS attack detection. This method was implemented over a NOX-based network and uses SOM algorithm to detect anomaly. The results indicated that their method was very effective and had high accuracy. In 2011, Syed Akbar Mehdi et al. [18] presented anomaly detection on SDN infrastructure. In this paper, many anomaly detection algorithms were used to validate that these methods were suitable for Small Office/Home Office (SOHO) environments. But these methods could only be applied to low network traffic rates. K. Gliotis et al. [4] combined sFlow [2] and Openflow [3] for anomaly detection. In this paper, authors compared sFlow method with native Openflow method for data collection. The experiment result indicated that sFlow was more efficient than native Openflow.

However, all above-mentioned methods only consider single network domain. While expanding to multi-domain, if multiple domains could cooperate with each other, the accuracy of anomaly detection would be improved. Soule et al. [7] showed that more anomalies were detected by analyzing the data of peering domains together. However, traffic data contains a lot of sensitive value, such as users' privacy, data flow information and so on. Hence, some of domains are unwilling to cooperate with other domains due to privacy disclosure risk.

To guarantee privacy, two major different methods are available which are cryptography method and geometric transformation method. The cryptography method has been widely studied in the literature and the most widely used one is secure multi-party computation. The secure multi-party computation tends to compute functions over inputs provided by multiple recipients without actually sharing the inputs with one another. The weakness of this method is its high execution time even though it guarantees the privacy of confidential data. There are many solutions with cryptography methods dealing with anomaly detection in legacy networks [12–14]. These solutions need tens to hundreds of seconds to detect anomaly which is unacceptable in actual environment.

Geometric transformation method has an extensive usage in statistical disclosure control due to its simplicity, efficiency, and ability to preserve statistical information [9–11]. The general idea is to replace the original data values with some synthetic ones. There are two methods in common use, data perturbation and data transformation. Data perturbation method adds some noises to original data to protect privacy. It is clear that such method reduces the accuracy of data usage even though preserves the privacy. Data transformation method converts original data to transformed data by orthogonal transformation or projection transformation so that the statistical data computed from the transformed data does not differ significantly from the statistical information computed from the original data. However, [15] shows that this kind of distance-preserving transformation method will breach privacy. Another drawback of geometric transformation method is that these methods unify the transformation functions or perturbation value which is infeasible in our scenario. If we could eradicate perturbation value in subsequent computation, we can use the data perturbation method to preserve high efficiency without sacrificing accuracy.

3 System Design

In SDN environment, network operators manage their domain by themselves due to the open network programmability of SDN. Every domain is managed by one or more controllers. There is no doubt that the detection rate will be higher with a larger traffic data set and more different viewpoints of the network. But traffic data contains a lot of sensitive value, such as users' privacy, data flow information etc.

- Users' privacy: Traffic data often contains source IP, destination IP, data flow start time and so on. These data will disclose users' surfing habits and associated users which belong to users' privacy.
- Data flow information: Traffic data often contains the number of packets per flow, the number of bytes per flow and so on. Disclosing data flow information will cause direct business attack.

In our paper, we use the SOM-based anomaly detection method to detect abnormal traffic. This method often takes as input aforementioned data. If we just use this method in multi-domain scenario, we should share these traffic data which will disclose privacy. Hence, most of the domains choose to abstract their domain as a virtual node, upload these coarse-grained information [1,8] and are unwilling to reveal the details of the intra information due to privacy consideration. To address this problem, we design an anomaly detection scheme which supports privacy protection.

3.1 Adversary Model

In our work, we assume that adversaries are semi-honest (or refer to honest but curious adversary) and adversaries can be any participants. That is to say, the participants are curious and attempt to learn from the information received by them, but do not deviate from the scheme themselves. By assuming a semi-honest threat model, we do not consider the scenarios where either the service provider or any of data providers is malicious. Malicious adversaries may do anything to infer secret information even aborting the scheme at any time. In our case, semi-honest model may be considered as more realistic adversary model because any of participants wants to obtain right results. The design of our scheme allows the collusion between data providers (e.g. controllers) while we do not allow the collusion between the service providers. The data providers can only collude with one of the service providers due to preceding constraint.

3.2 Design Goals

Under this adversary model, we define design goals as follows:

Privacy: We define privacy as follows: The participants can't learn the data held by other participants. Any other parties also should not be able to obtain the data of participants when doing some computation.

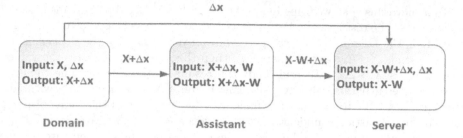

Fig. 1. Disturbance Erasion Process

Collaboration Benefits: The accuracy of anomaly detection across multiple domains should be higher than by single domain.

Efficiency: Our scheme should preserve real-time performance with privacy consideration.

Scalability: The overhead should not increase too fiercely with varying number of participant recipients.

Hence, our primary goals are privacy preserving, better accuracy, high efficiency and good scalability.

3.3 System Model

To achieve the goals mentioned above, we design an effective and privacy-preserving anomaly detection scheme across multi-domain. In this paper, we focus on privacy protection, so we use the existing SOM-based anomaly detection method. In order to gain the final model of SOM, we circularly train and adapt synaptic weights until SOM model has no obvious change. This method takes as input traffic data which will be protected carefully. We should obtain final adapted weights without knowing the traffic data. In this paper, we use the notation W to represent the synaptic weights of neuron vector and use the notation X to represent the traffic data.

To reduce communication overhead and computation cost, we use centralized system model in this paper, that is, we bring in a semi-honest third party, called server, to simplify the computation. The server uses traffic data collected from every domain to detect anomaly. As we stated above, semi-honest party executes scheme honestly but is curious about privacy of every domain. Hence, every domain can't send plaintext of traffic data to the server.

Every domain adds some random noise to their traffic data before sending in order to protect these data. As we mentioned above, data perturbation will cause accuracy sacrifice and every domain should unify noise value which is not realistic in our scenario. If we could erase noise in subsequent computation, we can use the data perturbation method to solve our problem without sacrificing accuracy.

To achieve this goal, we bring in another semi-honest third party, called assistant, to erase the disturbance. Disturbance erasion process is as shown in Fig. 1. First of all, the *Domain* shares the perturbation value Δx with the *Server*. The *Domain* sends $X + \Delta x$ to the *Assistant*, then the *Assistant* initiates the value W, calculates $X + \Delta x - W$ and sends the results to the *Server*. The *Server* knows the perturbation value Δx, she can compute $X - W$ accurately without knowing X and W. Reviewing the whole process, we can find that none of parties learn other parties' confidential data. However, data can be eavesdropped on the channel in real networks. The *Server* can eavesdrop the intermediate result $X + \Delta x$, then she can learn the precise data of the *Domain*. Hence, each party should encrypt its data before sending to other party.

To achieve efficiency goal, we utilize *Digital Envelope* to encrypt confidential data. *Digital Envelope* is a mechanism which combine asymmetric encryption method with symmetric encryption method. A sender uses symmetric secret key to encrypt sensitive data and then uses asymmetric public key to encrypt symmetric secret key. The sender sends encrypted data and encrypted symmetric key to the receipt. When the receipt receives these information, he uses asymmetric private key to decrypt the symmetric secret key. Then he uses this secret key to obtain plaintext. In this paper, we use *Digital Envelope* to combine RSA cryptography method and AES method. We can see this mechanism has ideal performance in the experiment part.

Now we describe the privacy preserving anomaly detection system overview using aforementioned design. To simplify the description, we assume that every domain is controlled by one controller in this paper. We consider a "semi-centralized" architecture which consists of several controllers, $C = \{C_i \mid i = 1, 2, \ldots, n\}$, a server S, and an assistant A as can be seen in Fig. 2. Every controller C_i collects flow statistics, such as packet counts, byte counts and so on. These controllers transform their sensitive data by adding a random noise, encrypt this transformed data by using *Digital Envelope* and contribute the processed data records to the assistant A which assists the server to erase the disturbance. Then these controllers share the perturbation value with the server S. The assistant collects the transformed traffic data from all domains and does some collaborative computation with the server.

After aforementioned preparation phase, the assistant holds the weights of neuron vector W while the server holds the difference values between the traffic data and weights $X - W$. The server cooperates with the assistant to adapt the synaptic weights. As we can see from Eq. 1, the server can compute the right part of Eq. 1.

$$W_j(t + 1) = W_j(t) + \eta(t)\Theta_j(t)(X(t) - W_j(t)); \tag{1}$$

The common parameter $\eta(t)$ and $\Theta_j(t)$ can also be calculated by the assistant, that is, these two parameters can't be used to protect the difference value $X - W$ and traffic data X. To protect these two value, we don't send the intermediate results to the assistant, but send the final results to the assistant. The detail of computation process is as follows:

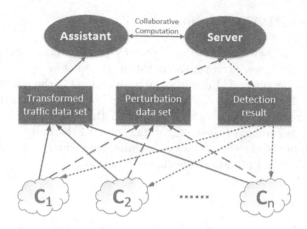

Fig. 2. System Architecture

1. When $t = 1$:
   The server uses the difference between sample of traffic data and initial weight $X(1) - W_j(0)$ to compute the difference between sample of traffic data and adapted weight $X(1) - W_j(1)$ with Eq. 2.

   $$X(1) - W_j(1) = (X(1) - W_j(0)) + \eta(1)\Theta_j(1)(X(1) - W_j(0)); \qquad (2)$$

   The server stores the intermediate results $X(1) - W_j(0)$ and $X(1) - W_j(1)$ for next training.

2. When other condition:
   The assistant can provide the difference value $X(t+1) - W_j(0)$ and the server uses Eq. 3 to compute $X(t+1) - W_j(t)$.

   $$X(t+1) - W_j(t) = (X(t+1) - W_j(0)) + (X(t) - W_j(t)) - (X(t) - W_j(0)); \qquad (3)$$

   The server uses Eq. 4 to adapt synaptic weights $X(t+1) - W_j(t+1)$.

   $$X(t+1) - W_j(t+1) = (X(t+1) - W_j(t)) + \eta(t)\Theta_j(t)(X(t+1) - W_j(t)); \qquad (4)$$

   The server stores the intermediate results $X(t+1) - W_j(0)$ and $X(t+1) - W_j(t+1)$ for next training.

3. Repeat the aforementioned step until that there is no obvious change in SOM model.
   The server obtains the intermediate results $X(T) - W_j(T)$, where T is the maximum training times. The server adds the relevant perturbation value and sends to the assistant. Because the assistant knows the transformed traffic data, it can computes the final adapted weights $X(T) - W_j(T)$. After adaption phase, the assistant holds final adaption weights while the server learns nothing.

4   Proposed PPAD Scheme

In this section, we will describe the privacy-preserving anomaly detection ($PPAD$). First of all, we interpret the notation used below.

S denotes the server and A denotes the assistant. D is the set of traffic data collected from all participant controllers. Δd is the set of perturbation value created by controllers. n denotes the number of the participant controllers. X represents a single sample of the traffic data. l means the number of neurons. c denotes the number of classes. $\eta(t)$ is the learning rate which gradually decreases with time t while $\Theta_j(t)$ is the neighborhood function.

4.1   Scheme Overview

We propose a scheme, named $PPAD$, which allows multiple domains to detect anomaly together without privacy disclosure. In our scheme, we use an SOM algorithm to detect anomaly. SOM is an artificial neural network trained with features of the traffic flow. It can be used to cluster existing traffic statistics to several classes which are used to detect new abnormal traffic. This scheme has two phases which are the training phase and the detection phase. To simplify description, we use typical encryption method instead of Digital Envelope encryption method in next section. The basic steps of our scheme are as follows:

Training phase:

1. All controllers transform their traffic data to two types and encrypt them using two secret keys.
2. The controllers send the processed data to the server and the assistant.
3. The server and the assistant cooperate to train the traffic data using SOM method while protecting domains' privacy.
4. The server and the assistant cooperate to classify the traffic data into two classes.

Detection phase:

1. All controllers use the same way described above to submit their processed data to the server and the assistant.
2. The server and the assistant classify the new flows into two classes collaboratively. If the new flow is classified into abnormal class, it is predicated as anomaly traffic and vice versa.
3. Return the final detection results to all controllers.

4.2   Training Algorithm

In the training phase, the goal is to compute the adapted weights and classes using existing traffic statistics collected from several controllers. Every controller holds the traffic data of their own domain. All controllers extract the characteristics of these data, transform the characteristics information and send them to the third party. Hence, they will not leak any information to the third party.

Algorithm 1 presents the privacy-preserving training algorithm. As we can see from the Algorithm 1, the server and the assistant can compute with plaintext of the data. To be simple, we call $D + \Delta d$ as transformed traffic data set while call Δd as perturbation data set. The assistant A chooses a single sample (i.e. $X + \Delta x$) from transformed traffic data set. Then he computes $X(t+1) + \Delta x(t+1) - W(0)$ and sends it to the server (line 10 in Algorithm 1). The server uses existing value 3 to compute $X(t+1) - W_j(t)$ with Eq. 3 (line 12 in Algorithm 1). After finding winning neuron, the server uses Eq. 4 to compute $X(t+1) - W_j(t+1)$ (line 16 in Algorithm 1). Repeat steps mentioned above until no significant change happens. Finally, the assistant A gets the adjusted weights $W(T)$ and the server S learns nothing.

Algorithm 1. Privacy-preserving Training Algorithm

Require:
   The traffic data D and the perturbation data Δd;
Ensure:
   The adjusted neuron vectors;
 1: A initializes W_j;
 2: **for** each controller C_i, $i \in [1, n]$ **do**
 3:     $C_i \rightarrow A : Enc_{ska}(D_i + \Delta d_i)$;
 4:     $C_i \rightarrow S : Enc_{sks}(\Delta d_i)$;
 5: **end for**
 6: $A : D + \Delta d = \bigcup_{i=1}^{n} Dec_{ska}(Enc_{ska}(D_i + \Delta d_i))$;
 7: $S : \Delta d = \bigcup_{i=1}^{n} Dec_{sks}(Enc_{sks}(\Delta d_i))$
 8: **while** t $\leq$ T and there is significant change in parameters **do**
 9:     A chooses a single sample $X(t+1) + \Delta x(t+1)$ from $D + \Delta d$;
10:     A computes $tmp = X(t+1) + \Delta x(t+1) - W(0)$ and sends it to S;
11:     **for** each $j \in [1, l]$ **do**
12:         S computes $X(t+1) - W_j(t) = (X(t+1) - W_j(0)) + (X(t) - W_j(t)) - (X(t) - W_j(0))$;
13:     **end for**
14:     S computes winning neuron $i(x) = \arg\min_j \|X(t+1) - W_j(t)\|$;
15:     **for** each $j \in [1, l]$ **do**
16:         S computes $X(t+1) - W_j(t+1) = (X(t+1) - W_j(t)) + \eta(t)\Theta_j(t)(X(t+1) - W_j(t))$;
17:     **end for**
18: **end while**
19: **for** each $j \in [1, l]$ **do**
20:     $S \rightarrow A : mid - res_j = X(T) - W_j(T) + \Delta x(T)$;
21: **end for**
22: A computes final weights of neuron vectors $W(T)$: $W(T) = mid - res_j - (X(T) + \Delta x(T))$;

4.3 Detection Algorithm

In the detection phase, the goal is to decide whether a new traffic set is abnormal or not. Similarly, every controller extracts the characteristics of their domains traffic data, transforms the characteristics information and sends them to the server S and the assistant A. Finally, the server returns the detection results to all domains. As we can see, the whole process does not leak any information to the third party.

Algorithm 2 presents the privacy-preserving detection algorithm. Initiation process is similar with training phase. If the traffic is classified into abnormal class, it is determined as an attack and vice versa (line 10 to 14 in Algorithm 2).

Algorithm 2. Privacy-preserving Detection Algorithm

Require:
    The traffic data Y and the perturbation data Δy;
    Normal class and Abnormal class;

Ensure:
    The detection result;

1: **for** each controller C_i, $i \in [1, n]$ **do**
2:    $C_i \rightarrow A : Enc_{ska}(Y_i + \Delta y_i)$;
3:    $C_i \rightarrow S : Enc_{sks}(\Delta y_i)$;
4: **end for**
5: $A : Y + \Delta y = \bigcup\limits_{i=1}^{n} Dec_{ska}(Enc_{ska}(Y_i + \Delta y_i))$;
6: $S : \Delta y = \bigcup\limits_{i=1}^{n} Dec_{sks}(Enc_{sks}(\Delta y_i))$
7: **for** each instance $Z + \Delta z$ in $Y + \Delta y$ **do**
8:    A computes $Z + \Delta z - W(T)$ and sends it to S;
9:    S classifies the new traffic
10:    **if** the traffic is classified into normal class **then**
11:      Z is normal traffic.
12:    **else**
13:      Z is abnormal traffic.
14:    **end if**
15: **end for**

---

5 Performance Evaluation

Seen from the whole training and detection process, all information is encrypted, so that the privacy of every party is preserved. Hence, we focus on performance evaluation in this section. We use SOM-based anomaly detection approach, where we cluster existing traffic statistics and do the training off-line and detect the real-time traffic online. In this case, we should do as much computation as possible off-line to improve the efficiency of online. We set neurons as a 40*40 matrix, initial learning rate as 0.5 and initial neighborhood radius as

20. We used Gaussian function to compute the neighborhood as done in the literature [17]. The servers operation was conducted on a 3.00 GHz Intel Pentium processor with 4GB memory, while the assistants operation was processed on a 2.13 GHz Intel Pentium processor with 3GB memory.

Collaboration Benefits. Soule et al. [7] showed that more anomalies were detected by analyzing the data of peering domains together. The author pointed out that anomaly detection performance will be different with diversity threshold and different kinds of traffic matrix. In this part, we will verify that if each domain could cooperate with each other, whether anomaly detection could be dealt with much higher accuracy or not in this scenario. We focus on the privacy issue, so that we just detect the typical attack-DDoS attack in this paper. Certainly, our scheme can be extended to detect other attack. We use 4-tuples which are Average of Packets per flow, Average of Bytes per flow, Average of Duration per flow and Growth of Different Ports respectively, more details can be seen in literature [5]. We use the Detection Rate (DR) to evaluate the performance of each scenario. The Detection Rate can be calculated with Eq. 5.

$$DR = \frac{TP}{TP + FN} \tag{5}$$

where TP (True Positives) mean anomalous traffic detected as attack, and FN (False Negatives) represent anomalous traffic classified as normal.

We will use six domains to do the validation test. Table 1 shows these different kinds of flows collected from several domains. For example, in Domain C_1, the controller collected 44,005 flows in total during the training phase, where 8,812 flows are normal and 35,193 flows are abnormal. During the detection phase, the controller from Domain C_1 collected 8,187 normal flows and 25,837 abnormal flows. The last row denotes the total number of the flows which are collected by the server. The server S and the assistant A cooperate to classify these data.

Table 2 shows the detection rate of each domain. As seen from table, when each domain detects anomalous traffic separately, the average of the detection

Table 1. The number of flows collected by each domain

Domain	Training phase		Detection phase	
	attack traffic	normal traffic	attack traffic	normal traffic
Domain C_1	35193	8812	25837	8187
Domain C_2	117643	28986	82563	22922
Domain C_3	58546	14757	38528	10317
Domain C_4	46953	11695	29368	9337
Domain C_5	78423	19347	61640	11139
Domain C_6	54700	13680	73800	8793
Server	391458	97277	306844	70695

Table 2. Detection results

Party	Domain C_1	Domain C_2	Domain C_3	Domain C_4	Domain C_5	Domain C_6	Server
DR (%)	86.7	94.33	88.6	92.7	91.11	90.94	98.95

rate is about 91 %. While the server gathers all traffic data to detect anomaly, the detection rate is about 99 %, which is much higher than detected by each domain.

Efficiency. Due to the privacy protection goal, our scheme requires extra computation and communication costs. Note that off-line computation and communication costs are not critical for overall performance. Hence, in this part, we focus on online performance.

We performed trials to find out how execution time changes with varying the number of participant domains. We performed experiments with different number of participant domains. The number of parties' ranges from 1 to 6. Each domain collects about 8000 flows and sends them to the third parties. In Fig. 3(a), we figure out the extra overhead of decryption operation on the server side and on the assistant side. As seen from Fig. 3(a), our cryptographical method doesn't cost much more extra overhead.

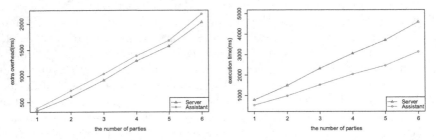

(a) Extra Overhead of Cryptographical Operation.   (b) Execution Time in two Service Providers.

Fig. 3. Performance Evaluation.

Scalability. Fig. 3(b) shows the execution time change with varying number of participant domains. In this experiment, each domain collects about 8000 flows and sends them to the third parties. Obviously, the overhead will go up with increasing participant parties because the server and the assistant should deal with much more data. Our main goal is to guarantee approximate linear growth. Note that the server and the assistant cooperate to detect the anomaly. The server should wait until the assistant complete some operation so that the server has higher execution time as seen from Fig. 3(b). The result shows that our scheme has tolerable overhead. The execution time of the scheme doesn't increase fiercely with increasing number of parties and has approximate linear growth which proves that our scheme can support scalability well.

6   Conclusion

In this paper, we first focused on the problem about multi-domain privacy protection for Software Defined Networks. To address this issue, we presented a privacy-preserving scheme to detect anomaly based on distributed data among multiple parties using artificial neural network algorithm. We focus on privacy, efficiency, accuracy and scalability. Since some of them are conflicting goals, we find a tradeoff scheme to solve this problem. We presented the design, deployment and evaluation of a collaborative anomaly detection scheme that is not only privacy preserving and have good performance. Moreover, experiments with our scheme implementation shows that our scheme works well with increasing numbers of parties.

As mentioned above, our scheme allows the collusion between data providers (e.g. controllers) and collusion between data providers and service providers (e.g. server or assistant). But we do not allow the collusion between service providers. As part of future work, we will continue to study and design a new method which can support collusion between service providers with high efficiency.

Acknowledgment. The research work reported in this paper is supported by National Science Foundation of China under Grant No. 61100172, 61272512, 61402037, Program for New Century Excellent Talents in University (NCET-12-0046), Beijing Natural Science Foundation No. 4132054, and Beijing Institute of Technology Research Fund Program for Young Scholars.

References

1. Koponen, T., Casado, M., Gude, N., et al.: Onix: a distributed control platform for large-scale production networks. In: OSDI, pp. 1–6 (2010)
2. Phaal, P.: sFlow Specification Version 5, July 2004
3. McKeown, N., Anderson, T., Balakrishnan, H., et al.: Openflow: enabling innovation in campus networks. ACM SIGCOMM Comput. Commun. Rev. **38**, 69–74 (2008)
4. Giotis, K., Argyropoulos, C., Androulidakis, G., et al.: Combining openflow and sflow for an effective and scalable anomaly detection and mitigation mechanism on SDN environments. Comput. Netw. **62**, 122–136 (2014)
5. Braga, R., Mota, E., Passito, A.: Lightweight DDoS flooding attack detection using NOX/OpenFlow. In: IEEE 35th Conference on Local Computer Networks (LCN), pp. 408–415. IEEE (2010)
6. Wang, B., et al.: DDoS attack protection in the era of cloud computing and Software-Defined Networking. Comput. Netw. **81**, 308–319 (2015)
7. Soule, A., Ringberg, H., Silveira, F., Rexford, J., Diot, C.: Detectability of traffic anomalies in two adjacent networks. In: Uhlig, S., Papagiannaki, K., Bonaventure, O. (eds.) PAM 2007. LNCS, vol. 4427, pp. 22–31. Springer, Heidelberg (2007)
8. Lin, P., Bi, J., Chen, Z., et al.: WE-bridge: West-East Bridge for SDN inter-domain network peering. In: IEEE Conference on Computer Communications Workshops (INFOCOM WKSHPS), pp. 111–112. IEEE (2014)

9. Oliveira, S.R.M., Zaiane, O.R.: Privacy preserving clustering by data transformation. J. Inf. Data Manag. **1**, 37 (2010)
10. Chen, K., Liu, L.: Privacy-preserving multiparty collaborative mining with geometric data perturbation. IEEE Trans. Parallel Distrib. Syst. **20**(12), 1764–1776 (2009)
11. Erfani, S.M., Law, Y.W., Karunasekara, S., Leckie, C.A., Palaniswami, M.: Privacy-preserving collaborative anomaly detection for participatory sensing. In: Ho, T.B., Zhou, Z.-H., Chen, A.L.P., Kao, H.-Y., Tseng, V.S. (eds.) PAKDD 2014, Part I. LNCS, vol. 8443, pp. 581–593. Springer, Heidelberg (2014)
12. Nagaraja, S., Jalaparti, V., Caesar, M., Borisov, N.: P3CA: private anomaly detection across ISP networks. In: Fischer-Hübner, S., Hopper, N. (eds.) PETS 2011. LNCS, vol. 6794, pp. 38–56. Springer, Heidelberg (2011)
13. Zhang, P., Huang, X., Sun, X., et al.: Privacy-preserving anomaly detection across multi-domain networks. In: 9th International Conference on Fuzzy Systems and Knowledge Discovery (FSKD), pp. 1066–1070. IEEE (2012)
14. Nguyen, H.X., Roughan, M.: Multi-observer privacy-preserving hidden Markov models. IEEE Trans. Signal Process. **61**, 6010–6019 (2013)
15. Giannella, C.R., Liu, K., Kargupta, H.: Breaching Euclidean distance-preserving data perturbation using few known inputs. Data Knowl. Eng. **83**, 93–110 (2013)
16. Lindell, Y., Pinkas, B.: Secure multiparty computation for privacy-preserving data mining. J. Priv. Confidentiality **1**, 59–98 (2009)
17. Lo, Z.P., Fujita, M., Bavarian, B.: Analysis of neighborhood interaction in Kohonen neural networks. In: 6th International Parallel Processing Symposium, CA, Los Alamitos (1991)
18. Mehdi, S.A., Khalid, J., Khayam, S.A.: Revisiting traffic anomaly detection using software defined networking. In: Sommer, R., Balzarotti, D., Maier, G. (eds.) RAID 2011. LNCS, vol. 6961, pp. 161–180. Springer, Heidelberg (2011)
19. Giotis, K., Androulidakis, G., Aglaris, V.: Leveraging SDN for efficient anomaly detection and mitigation on legacy networks. In: Third European Workshop on Software Defined Networks (EWSDN), pp. 85–90. IEEE (2014)
20. Chung, C.-J., Nice, et al.: Network intrusion detection and countermeasure selection in virtual network systems. IEEE Transactions on Dependable and Secure Computing, pp. 198–211 (2013)
21. IEEE SDN For. 2013, 1–7 (2013)
22. Kreutz, D., Ramos, F., Verissimo, P.: Towards secure and dependable software-defined networks. In: Proceedings of the second ACM SIGCOMM workshop on Hot topics in software defined networking, pp. 55–60 (2013)
23. Zhan, J.: Privacy-preserving collaborative data mining, Computational Intelligence Magazine, pp. 31–41. IEEE (2008)
24. Aggarwal, C.C., Philip, S.Y.: A general survey of privacy-preserving data mining models and algorithms. In: Aggarwal, C.C., Philip, S.Y. (eds.) A General Survey of Privacy-Preserving Data Mining Models and Algorithms. Advances in Database Systems, vol. 34, pp. 11–52. Springer, Heidelberg (2008)

Distributed Multi-user, Multi-key Searchable Encryptions Resilient Fault Tolerance

Huafei Zhu(✉)

School of Computer and Computing Science, Zhejaing University City College,
Hangzhou 310015, China
zhuhf@zucc.edu.cn

Abstract. In this paper, a construction of distributed multi-user, multi-key searchable encryptions is proposed and analyzed. Our scheme leverages a combination of the Shamir's threshold secret key sharing, the Pohlig-Hellman function and the ElGamal encryption scheme to provide high reliability with limited storage overhead. It achieves the semantic security in the context of the keyword hiding, the search token hiding and the data hiding under the joint assumptions that the decisional Diffie-Hellman problem is hard and the pseudo-random number generator deployed is cryptographically strong.

Keywords: ElGamal encryption scheme · Pohlig-Hellman function · Searchable encryptions · Shamir's threshold secret key sharing

1 Introduction

Often, different users possess data of mutual interest. The most challenging aspect of the data exchange lies in supporting of the data sharing over the encrypted database [12, 18, 28, 32]. Searchable encryption is such a cryptographic primitive allowing for the keyword based content sharing managed and maintained by individual users. The state of the art research on searchable encryptions can be classified as the following two categories:

– Different data items (or documents, we do not distinguish the notion of data with that of the document throughout the paper as two nations are interactively used in many references cited here) outsourced are encrypted by a single key. The private information retrieval line of work [4, 8, 11, 20] and the oblivious transfer line of work [25] fall in this category. Most of the research on searchable encryptions focused on the case when data is encrypted with same key [3, 5, 7, 10, 13, 15, 28, 29, 32] and more efficient solutions [1, 9, 19, 21, 24, 26, 27, 30, 31] have been proposed in recent years. The idea behind these constructions is that − to access a database, individually authorized user is issued a query key by the data owner and only the authorized users who have valid query keys can generate valid access queries which enable the database management server to process users' search queries without learning the keywords contained in the queries and the contents of the encrypted records.

© Springer International Publishing Switzerland 2016
M. Yung et al. (Eds.): INTRUST 2015, LNCS 9565, pp. 17–31, 2016.
DOI: 10.1007/978-3-319-31550-8_2

– Different data outsourced are encrypted with different keys. This problem was first studied by López-Alt et al. [17] leveraging the concept of fully homomorphic encryption schemes in which anyone can evaluate a function over data encrypted with different keys. The decryption requires all parties to come together and to run a multi-party computation protocol so that a client can retrieve all the keys under which the data was encrypted. As a result, the users need to do work proportional in the number of keys. Very recently, Popa and Zeldovich [22,23] proposed alternative design based on the bilinear map. Roughly speaking, a data owner in their model generates a set of documents $\{d_1, \ldots, d_\lambda\}$ and then an access structure is defined for all users. Each document d_j generated and the corresponding keyword set $\{w_j^{(1)}, \ldots, w_j^{(n_j)}\}$ extracted at session j $(sid = j)$ will be encrypted by a fresh secret key k_j. The encrypted data and keyword set are then outsourced to a database server. A legitimate user is then given the corresponding encryption key k_j via a secure and authenticated channel established between the participants.

1.1 The Motivation Problem

Note that in the Popa and Zeldovich's scheme [22,23], user's primary key is assigned by the data owner while the corresponding delta keys are computed from the primary key and the specified encryption keys. The underlying access graph should be updated whenever a new document is outsourced to the server. The update procedure could be a difficult task if the frequency of data outsourcing is high since the size of stored delta keys can be proportional to the stored documents. Furthermore, when the deployed server is unreliable, as the case in modern data centers, redundancy must be introduced into the system to improve reliability against server failure (say, a complicated delta key recovery mechanism, or a Hadoop-like delta key duplication mechanism or a MapReduce-like distributed computing mechanism should be introduced). Since no countermeasure dealing with the server failure (or the delta key recovery) is known within the multi-user, multi-key searchable encryption framework, it is certainly welcome if one is able to provide such a counter-measure resilient the server failure.

1.2 This Work

This paper studies multi-user, multi-key searchable encryptions in the data owner controlled framework, where a data owner generates, manages and maintains all generated documents, documents encryption keys and keyword encryption keys (we distinguish the keys used to encrypt documents and to encrypt keywords throughout the paper). In our model, a database management system $(DBMS)$, a data owner O, a set of users, a token generator (TG), a token extractor (TE) and a data extractor (DE) are introduced and formalized. All participants in our model run in the X-as-a-service model, where $X = $ (token generation, token extraction, data extraction etc.):

- The $DBMS$ manages and maintains the system level parameters in the bulletin board model. The $DBMS$ should be able to add public information to the bulletin board so that all participants are able to obtain public information from the bulletin board. We stress that bulletin boards are used in any instance where public access to information is desired in the cryptography.
- A data owner O generates his/her data in a session and then extracts a set of keywords from the generated data accordingly (for example, by means of the inverted index program). To outsource the data generated at the ith session (the session id is denoted by $sid = i$), O first generates a secret document encryption key sk_i that will be used to encrypt the document. Then a public mask key K_i that will be used to mask the document encryption key sk_i and a keyword encryption key k_i that will be used to encrypt the generated keyword set are generated by means of a cryptographically secure random number generator.

  Let $t_i = (t_i^{(1)}, t_i^{(2)})$ be an output of cryptographically strong sustainable pseudo-random number generator at the ith session (say, the Barak-Halevi's (BH) scheme [2], or any other cryptographically strong pseudo-random number generator). $t_i^{(1)}$ is used to generate the mask encryption key $K_i \leftarrow g^{t_i^{(1)}}$ while $t_i^{(2)}$ is used to generate the keyword encryption key $k_i = H(t_i^{(2)})$, where $< g >= G \subseteq Z_p^*$, $|G| = 2q$, $p = 2q + 1$ is a large prime number and H: $\{0,1\}^* \rightarrow G$, is a cryptographically strong hash function. The auxiliary mask encryption string $t_i^{(1)}$ is shared among a set of n_D data extraction processors where any subset of m_D-out-of-the-n_D processors can be used to reconstruct K_i while the auxiliary keyword encryption string $t_i^{(2)}$ is kept secret by the data owner. The encrypted data are then outsourced to the $DBMS$.
- To support the keyword search procedure, the data owner must provide search structures for users. In many real-life situations, we don't believe that any given person can be trusted, yet it is reasonable to assume that the majority of people are trustworthy. Similarly, in on-line transactions, we may doubt that a given server can be trusted, but we hope that the majority of servers are working properly. Based on this assumption, we can create trusted entities, where the notion of token generators which manage and maintain a set of token generation processors, the notion of token extractors which manage and maintain a set of token extractor processors and the notion of data extractors which manage and maintain a set of data extraction processors are introduced. All keyword encryption keys are securely shared among token generation processors while all auxiliary mask strings are securely shared among the data extraction processors by means of the Shamir's secret sharing protocol.

An Overview of Processing. A processing of a keyword search comprises the following phases: the setup phase (including the data outsourcing); the query processing phase and the data extraction phase.

- In the setup phase, system parameters are generated for all participants; The data owner generates document encryption keys, mask keys and keyword encryption keys for the initial data set. The auxiliary mask strings and keyword encryption keys are then securely distributed among a set of data extraction processors and a set of token generation processors respectively.

- In the query processing phase, a user first selects a keyword w, and then encrypts it by the ElGamal encryption scheme $(u = g^r, v = H(w)h^r)$, where g and H are common system strings and h is generated on-the-fly from an arbitrary subset of token extraction processors. The encrypted keyword $c = (u, v)$ is then sent to the $DBMS$ via the token generator server TG. The $DBMS$ and token extraction processors TEPs work together to extract the search token, and then retrieve data accordingly from the database server;
- In the data extraction phase, the retrieved ciphertexts such that each of which contains the specified keyword w are sent back to the user. The user then invokes m_D-out-of-n_D data extraction processors to decrypt the ciphertexts.

The Security. Intuitively, we expect the semantic security from multi-key searchable schemes:

- Keyword hiding: an adversary cannot learn the keyword one searches for;
- Token hiding: an adversary is not be able to distinguish between ciphertexts of two search tokens;
- Data hiding: if a document encryption key leaks, the contents of the other documents the user has access should not leak.

We are able to show that if the Diffie-Hellamn problem is hard and the underlying pseudo-random number generator is cryptographically strong, then the proposed multi-key searchable encryption is semantically secure.

What's New? We provide a new construction of multi-user, multi-key searchable encryptions based on the Pohlig-Hemman function and the ElGamal encryption scheme and a new method of achieving on-the-fly multi-party computation using the threshold multi-key encryptions. Our solution is different from the state-of-the-art solutions [17] leveraging the lattice based encryption scheme NTRU [14]), where a-priori bounded number of users should be defined since a decryption depends on the specified bound. Our solution is also different from the Popa and Zeldovich's methodology [22,23] which is leveraging the bilinear map based encryption scheme [6]), where a document encryption key should be distributed to all valid users. The proposed scheme leverages a combination of the Shamir's threshold secret key sharing, the Pohlig-Hellman function and the ElGamal encryption scheme to provide high reliability with limited storage overhead. It achieves the semantic security (the keyword hiding, the search token hiding and the data hiding) under the joint assumptions that the decisional Diffie-Hellman problem is hard and the pseudo-random number generator deployed is cryptographically strong.

The Road Map: The rest of this paper is organized as follows: in Sect. 2, syntax and security definition of multi-user, multi-key search protocols are presented;

An efficient construction of searchable encryptions based on the Pohlig-Hellman functions and the ElGamal encryption scheme is then proposed and analyzed; We show the proposed scheme is semantically secure in Sect. 3. We conclude this work in Sect. 4.

2   Syntax and Security Definition

This section consists of the following two parts: syntax and security definitions of multi-user, multi-key searchable encryptions.

2.1   Syntax of Multi-user, Multi-key Database Search

A multi-user, multi-key database search scheme comprises the following participants: a database management system, data owners, users, a token generator, a token extractor and a data extractor. Our scheme works in the bulletin board model where a participant can add his/her public information to it so that any other participant can use the public information available on the bulletin board. Notice that once the public information is outsourced to the bulletin board it cannot be deleted or modified by the original public information creator. The integrate of the outsourced public information is managed and maintained by a trusted certificate authority.

1. A database management system ($DBMS$) takes as input the security parameter 1^k and outputs system wide parameters **params** and a pair of public/secret keys (pk_{DB}, sk_{DB}). **params** is publicly known by all participants.
2. A set of data owners are involved in a searchable encryption scheme. Each data owner (O) takes as input the system parameters **params** and outputs a pair of public and secret keys (pk_O, sk_O);

   A procedure for outsourcing the encrypted data will be modelled as a session. In each session $sid = j$, O takes **params** and (pk_O, sk_O) as input and generates a triple of the document encryption key $sk_j^{(O)}$, the mask encryption key $K_j^{(O)}$ and the keyword encryption key $k_j^{(O)}$;
3. A token generator TG takes **params** as input and generates a pair of its own public and secret keys (pk_{TG}, sk_{TG}).

   A token generator manages and maintains a group of token generation processors (TGPs). Each TGP takes **params** as input and generates a pair of public and secret keys (pk_{TGP_i}, sk_{TGP_i}). To enable the keyword search over the outsourced encrypted data, O will distribute secret shares of a keyword encryption key $k_j^{(O)}$ generated at session j to n_G token generation processors by means of the Shamir's threshold secret key sharing scheme such that any subset of m_G (out-of-n_G) token generation processors can reconstruct the keyword encryption key $k_j^{(O)}$.
4. A token extractor (TE) takes **params** as input and generates a pair of its own public and secret keys (pk_{TE}, sk_{TE}). TE manages and maintains a group of token extraction processors (TEPs). Each TEP takes **params** as input and generates a pair of public and secret keys (pk_{TEP_i}, sk_{TEP_i}) for $i = 1, \ldots, n_E$.

5. A data extractor DE takes params as input to generate a pair of public and secret keys (pk_{DE}, sk_{DE}). DE manages and maintains a group of n_D data extraction processors $(DEP_1, \cdots, DEP_{n_D})$. Each data extraction processor DEP takes params and pk_{DE} as input to generate a pair of public and secret keys (pk_{DEP_i}, sk_{DEP_i}); To enable users to extract the retrieved encrypted documents, O will distribute shares of the auxiliary mask key string of $K_j^{(O)}$ generated at session j to n_D data extraction processors by means of the Shamir's threshold secret key sharing scheme such that any subset of m_D (out of n_D) data extraction processors can reconstruct the mask encryption key $K_j^{(O)}$.

6. A set of users are involved in the searchable encryption. Each user U (or an querier) takes params as input to generate a pair of public and secret keys (pk_U, sk_U). A valid user is allowed to submit a query to the $DBMS$. On input a keyword $w \in W$, U encodes w and then sends the resulting codeword $c(w)$ to TG who generates a valid search token $t(w)$ by means of the multi-party computations. The resulting search token $t(w)$ is then sent to the $DBMS$ who collaborates with TE to extract the search token and then sends back to U all retrieved encrypted data $\mathcal{D}$ such that each $D \in \mathcal{D}$ contains w.

2.2   Security of Multi-key Database Search

We formalize security requirements specified in Sect. 1 with following games: keyword hiding, token hiding and data hiding that express these goals. One holistic security definition would be a stronger guarantee, but that greatly complicate the designs and proofs. Nevertheless, the separate definitions also capture the desired security goals.

Keyword Hiding Game. The keyword hiding game is between a challenger $\mathcal{C}$ and an adversary $\mathcal{A}$ on security parameter 1^k and pubic parameter params

- $\mathcal{C}$ invokes the $DBMS$ which takes as input 1^k to output params and provides params to $\mathcal{A}$;
- $\mathcal{C}$ invokes n token extraction processors each of which takes as input params to output n pairs of public and secret keys (pk_{TEP_i}, sk_{TEP_i}) $(i = 1, \cdots, n)$. $\mathcal{C}$ then provides pk_{TEP_i} $(i = 1, \cdots, n)$ to $\mathcal{A}$.
- Let Δ be an arbitrary subset of $\{1, \cdots, n\}$ containing m public key indexes. Let $pk_{TEP}^{(\Delta)}$ be a public key computed from the selected m public keys.
- Let w_0 and w_1 be two keywords selected by $\mathcal{A}$. The challenger selects a bit $b \in_R \{0, 1\}$ uniformly at random. Let $c = E_{pk_{TEP}^{(\Delta)}}(w_b)$ for $b \in_R \{0, 1\}$. The adversary is given (Δ, c) and outputs a guess $b' \in \{0, 1\}$.

Definition 1 (*Keyword Hiding*). *We say that the communication between users and the token extractor (and the token extraction processors) is keyword hiding if for any polynomial time adversary $\mathcal{A}$ that given (Δ, c), where $c = E_{pk_{TEP}^{(\Delta)}}(w_b)$ for $b \in_R \{0, 1\}$, outputs a guess b', the following holds: $Pr[b = b'] - 1/2$ is at most a negligible amount:*

Token Hiding Game. The token hiding game is between a challenger $\mathcal{C}$ and an adversary $\mathcal{A}$ on security parameter 1^k and pubic parameter params

- $\mathcal{C}$ invokes the $DBMS$ which takes as input 1^k to output params and provides params to $\mathcal{A}$;
- $\mathcal{C}$ invokes the data owner O which takes as input the system parameters params to output a pair of public and secret keys (pk_O, sk_O). The adversary $\mathcal{A}$ is given pk_O; In each session, say session j, O takes params as input and generates a keyword encryption key k_j;
- The adversary $\mathcal{A}$ invokes the database management server which takes as input params to output (pk_{DB}, sk_{DB}). $\mathcal{A}$ obtains pk_{DB};
- $\mathcal{C}$ invokes n token generation processors each of which takes as input params to output n pairs of public and secret keys (pk_{TGP_i}, sk_{TGP_i}) $(i = 1, \cdots, n)$. $\mathcal{C}$ then provides pk_{TGP_i} $(i = 1, \cdots, n)$ to $\mathcal{A}$;
- Let $k_j^{(l)}$ be a secret share of k_j shared by TGP_l for $j = 1, \cdots, \kappa$ and $l = 1, \cdots, n$, where κ is the number of k_j shared among the token processors so far. Let w be an input to the token generation processor TGP_j. The challenger then invokes TGP_j which takes $k_j^{(l)}$ as an input and then outputs $c_j^{(l)}$. The resulting ciphertext $c_j^{(l)}$ is then sent to $\mathcal{A}$ who computes the corresponding coefficient $\alpha_j^{(l)}$ of the Lagrange Interpolation Formula to output an encryption $TG(k_j, w)$ of the search token.
- The challenger then selects a bit $b \in \{0, 1\}$ uniformly at random and then given $\{c_j^{(l)}\}_{l=1}^n$ and $TG_1 = (TG(k_j, w))$ if $b = 1$ and $\{c_j^{(l)}\}_{l=1}^n$ and a random string $TG_0 \in G$ if $b = 0$. The adversary outputs a guess $b' \in \{0, 1\}$.

Definition 2 *(Token Hiding). We say that the communication between users and the token generator who manages and maintains the token generation processors is token hiding if for any polynomial time adversary $\mathcal{A}$ that given $TG_b(k, c)$ and $\{c_j^{(l)}\}_{l=1}^n$, outputs a guess b', the following holds: $Pr[b = b'] - 1/2|$ is at most a negligible amount:*

Data Hiding Game. The data hiding game is between a challenger $\mathcal{C}$ and an adversary $\mathcal{A}$ on security parameter 1^k and pubic parameter params

- $\mathcal{C}$ invokes the $DBMS$ which takes as input 1^k to output params and (pk_{DB}, sk_{DB}). $\mathcal{C}$ provides params and pk_{DB} to $\mathcal{A}$;
- $\mathcal{C}$ invokes the data owner O which takes as input the system parameters params to output a pair of public and secret keys (pk_O, sk_O). The adversary $\mathcal{A}$ is given pk_O;

  In each session, say session j, O takes params as input and generates a mask encryption key K_j. The data m_j is then encrypted under K_j. The resulting ciphertext c_j is outsourced to the $DBMS$;
- $\mathcal{C}$ invokes n data extraction processors each of which takes as input params to output n pairs of public and secret keys (pk_{DEP_i}, sk_{DEP_i}) $(i = 1, \cdots, n)$. $\mathcal{C}$ then provides pk_{DEP_i} $(i = 1, \cdots, n)$ to $\mathcal{A}$;

Let $K_j^{(l)}$ be the share of auxiliary mask key string of K_j by DEP_l for $j = 1, \cdots, \kappa$ and $l = 1, \cdots, n$, where κ is the number of K_j shared among the data extraction processors so far.

To decrypt a ciphertext c_j, the data extractor first selects an arbitrary subset of $\{1, \cdots, n\}$ that contains arbitrary m public key indexes of the data extraction processors. Let Δ be the selected subset. The data extractor then invokes $DEP_j \in \Delta$ which takes $K_j^{(l)}$ as an input to output $c_j^{(l)}$. The resulting $c_j^{(l)}$ is then sent back to DE who computes the corresponding coefficient $\alpha_j^{(l)}$ of the Lagrange Interpolation Formula to output the plaintext m.

– Let m be a target document selected by $\mathcal{A}$. For the given m, $\mathcal{C}$ selects a random bit $b \in_R \{0,1\}$. Let $c_b = E_{DEP}(m)$ for $b \in_R \{0,1\}$ and $c_{\bar{b}} = E_{DEP}(1^{|m|})$ (an encryption of the dummy document). The adversary is given (c_0, c_1), and outputs a guess $b' \in \{0,1\}$.

Definition 3 *(Data Hiding). We say that the communication between the owner and the data extractor is data hiding if for any polynomial time adversary $\mathcal{A}$ that given (c_0, c_1), outputs a guess b', the following holds: $Pr[b = b'] - 1/2|$ is at most a negligible amount:*

Definition 4 *(Semantic Security). We say that a multi-user, multi-key searchable encryption system is semantically secure if it achieves the keyword hiding, token hiding and data hiding properties.*

3   The Construction

We now present a construction of multi-user, multi-key searchable encryptions in the bulletin board model that realizes the functionalities described in Sect. 3.1. We analyze its security in Sect. 3.2.

3.1   The Description

Our protocol comprises the following phases: the setup phase, the outsourcing phase, the processing phase and the extraction phase. The details of each phase are depicted below

The setup phase

– On input a security parameter parameter 1^k, $DBMS$ output system parameters **params**: a large safe prime number p such that $p = 2q + 1$, p and q are prime numbers, $|p| = k$ together with a cyclic group G of order q. Let g be a random generator of G.

$DBMS$ then takes **params** as input to generate a pair of public and secret keys (pk_{DB}, sk_{DB}), where $pk_{DB} = (g, h_{DB})$, $h_{DB} = g^{x_{DB}}$ and $sk_{DB} = x_{DB}$.

– A data owner O takes **params** as input and generates a pair of public and secret keys (pk_O, sk_O) where $pk_O = (g, h_O)$, $h_O = g^{x_O} \bmod p$ and $sk_O = x_O$ (in the following discussions, we simply assume that (pk_O, sk_O) is suitable for both the data encryption and data attestation).

- A data extractor DE takes **params** as input to generate a pair of public and secret keys (pk_{DE}, sk_{DE}), where $pk_{DE} = (g, h_{DE})$, $h_{DE} = g^{x_{DE}}$ and $sk_{DE} = x_{DE}$.

    DE in our model manages and maintains n_D data extraction processors $(DEP_1, \cdots, DEP_{n_D})$. Each extraction processor DEP_i generates its own public and secret key pairs $pk_{DEP_i} = (g, h_{DEP_i})$, $h_{DEP_i} = g^{x_{DEP_i}}$ and $sk_{DEP_i} = x_{DEP_i}$ independently.

    To enable users to obtain the corresponding plaintexts from the retrieved encrypted data, O delegates her decryption right to data extraction processors by invoking the Shamir's (m_D, n_D)-secret-key sharing algorithm such that any m_D combinations of shares is sufficiently to reconstruct the mask encryption key by applying the Lagrange Interpolation Formula.

    For simplicity, we assume that a secure (private and authenticated) channel has been established between O and DE and secure channels between DE and DEP_i respectively (such a secure channel assumption can be eliminated trivially under the standard PKI assumption).
- A token generator TG takes **params** as input to output a pair of public and secret keys (pk_{TG}, sk_{TG}), where $pk_{TG} = (g, h_{TG})$, $h_{TG} = g^{x_{TG}}$ and $sk_{TG} = x_{TG}$.

    In our model, TG manages and maintains n_G token generation processors $(TGP_1, \cdots, TGP_{n_G})$. Each token generation processor TGP_i generates its own public and secret key pairs $pk_{TGP_i} = (g, h_{TGP_i})$, $h_{TGP_i} = g^{x_{TGP_i}}$ and $sk_{TGP_i} = x_{TGP_i}$. Again, we assume that a secure channel has been established between TG and O (TG and TGP_i respectively).
- A token extractor TE takes **params** as input to output a pair of public and secret keys (pk_{TE}, sk_{TE}), where $pk_{TE} = (g, h_{TE})$, $h_{TE} = g^{x_{TE}}$ and $sk_{TE} = x_{TE}$.

    In our model, TE manages and maintains n_E token extraction processors $(TEP_1, \cdots, TEP_{n_E})$. Each token extraction processor TEP_i generates its own public and secret key pairs $pk_{TEP_i} = (g, h_{TEP_i})$, $h_{TEP_i} = g^{x_{TEP_i}}$ and $sk_{TEP_i} = x_{TEP_i}$. We assume that a secure channel has been established between TE and $DBMS$ (TE and TEP_i respectively).
- A user U takes **params** as input to generate a pair of public and secret keys (pk_U, sk_U), where $pk_U = (g, h_U)$, $h_U = g^{x_U}$ and $sk_U = x_U$.

The outsourcing phase

In the outsourcing phase, the search structure of the outsourced data and keyword is defined as follows: let BH be the Barak-Helavi's (or any other cryptographically strong) pseudo-random number generator [2] and $H: \{0,1\}^* \to G$ be a cryptographically strong hash function. We view a data outsourcing activity as a session in the following depiction.

- At session $sid = i$, on input d_i, the data owner O first selects a document encryption key sk_i with suitable length (say 128-bit or 256-bit for AES) and then invokes the BH pseudo-random number generator to output a pair of mask key K_i and keyword encryption key k_i. The document encryption key sk_i is encrypted by the mask encryption key K_i computed on the fly: let s_{i-1} be the previous internal state of the BH pseudo-random number generator at session $sid = i - 1$. To generate a pair of mask key K_i and keyword encryption key k_i for d_i, O invokes the BH scheme which takes s_{i-1} as input to output (s_i, t_i), where s_i is the internal state at session $sid = i$ and t_i is the current output. O then parses t_i to two parts $(t_i^{(1)}, t_i^{(2)})$ and then enciphers the first part $t_i^{(1)}$ by computing $K_i = g^{t_i^{(1)}} \bmod p$ and the second part $t_i^{(2)}$ by computing $k_i = H(t_i^{(2)})$, where H is a cryptographically hash function (in essence, we view H as a random oracle). K_i is called the mask key that will be used to encrypt the document encryption key sk_i while k_i is called the keyword encryption key. The first part $t_i^{(1)}$ is called the auxiliary mask string while the second part $t_i^{(2)}$ is called the auxiliary keyword encryption string.

– O extracts the keyword sets $W_i = \{w_i^{(1)}, \cdots, w_i^{(\gamma_i)}\}$ from d_i by means of the inverted index.

To encrypt a keyword $w \in W_i$, O invokes the Pohlig-Hellman function to compute $c(k_i, w) \leftarrow H(w)^{k_i} \bmod p$. $c(k_i, w)$ is then outsourced the $DBMS$. Let $c(k_i, W_i)$ be an encryption of the keyword set W_i under the keyword encryption key k_i at session $sid = i$.

To enable users to search keywords, the data owner O provides a search structure by sharing k_i among n_G token generation processors managed and maintained by the token generator TG. To distribute secret shares to $TGPs$, O invokes the Shamir's threshold secret key sharing protocol below:

- O randomly selects a polynomial $f(x) = f_0 + f_1 x + \cdots + f_{m_G-1} x^{m_G-1}$ (mod q), where $f_0 = k_i$ and $k_i^{(l)} \overset{def}{=} f(TEP_l))$ $(i = 1, \ldots, n_G)$;
- TEP_l is given $f(TEP_l)$, $l = 1, \ldots, n_G$.

– To outsource d_i, O first invokes a cryptographically strong block cipher say, Advanced encryption standard AES which takes sk_i and d_i as input to generate the ciphertext $c(sk_i, d_i)$.

Let $c(K_i, sk_i) = (g^r, sk_i \times K_i^r)$ be an encryption of the secret key sk_i under the mask encryption key K_i. The corresponding auxiliary mask encryption string is shared among n_D data extraction processors $DEP_1, \ldots, DEP_{n_D}$ (again O applies the Shamir's threshold scheme to the auxiliary string $t_i^{(1)}$) such that a combination of m_D-out-of-n_D shares can be used to reconstruct the the the auxiliary key $t_i^{(1)}$ such that $K_i = g^{t_i^{(1)}}$.

The resulting ciphertext $(c(k_i, W_i), c(K_i, sk_i), c(sk_i, d_i))$ are then sent to the $DBMS$.

The query processing phase

In the query processing phase, a computation of individual user is depicted below

– The input of a user U is a keyword w together with a description of token extraction processors whose public keys are denoted by $(g, h_{TEP_1}), \cdots, (g,$

$h_{TEP_{n_E}}$). U then selects m_E-out-of-n_E token extraction processors uniformly at random. Let $(g, h_{TEP_{i_1}}), \cdots, (g, h_{TEP_{i_{m_E}}})$ be m_E selected token extraction processors. Let $\Delta = (i_1, \cdots, i_{m_E})$ and $h = \Pi_{j=1}^{m_E} h_{TEP_{i_j}}$.

To hide the selected keyword w, U selects a string $r \in Z_q$ uniformly at random and computes $u = g^r \bmod p$, $v = H(w)h^r \bmod p$. Let $c = (u, v)$ and $\widetilde{\Delta}$ be an encryption of Δ under the $DBMS$' public-key pk_{DB}, i.e., $\widetilde{\Delta} = E_{pk_{DB}}(\Delta)$. Let $m_U = (\widetilde{\Delta}, c)$. Let $\sigma_U(H(m_U)$ be a signature of m_U attested by the user U. $(\sigma_U(H(m_U)), m_U)$ is then sent by the user to the token generator.

- Upon receiving $(\sigma_U(H(m_U)), m_U)$, the token generator TG who manages n_G token generation processors first checks the validity of the received message (recall that all computations are running in the X-as-a-service model). If the signature is invalid, then TG rejects the received message; otherwise, TG selects m_G-out-of-the-n_G token generation processors uniformly at random and then forwards $c = (u, v)$ to the selected m_G processors $\{TGP_{i_1}, \cdots, TGP_{i_{m_G}}\}$. Let $k_j^{(l)}$ be a secret share of k_j by TGP_l for $j = 1, \cdots, \kappa$ and $l = 1, \cdots, n_G$, where κ is the number of k_j shared so far. For each share $k_j^{(l)}$, TGP_l performs the following computations for each share $k_j^{(l)}$:

  - $u_j^{(l)} = u^{k_j^{(l)}} \bmod p$;
  - $v_j^{(l)} = v^{k_j^{(l)}} \bmod p$.

  TGP_j then sends $c_j^{(l)}$ back to TG, where $c_j^{(l)} = (u_j^{(l)}, v_j^{(l)})$;

  Upon receiving $c_j^{(l)}$, TG computes the corresponding coefficient $\alpha_j^{(l)}$ of the Lagrange Interpolation Formula and then computes $u_j = \Pi_{l=i_1}^{i_{m_T}} (u_j^{(l)})^{\alpha_j^{(l)}}$ and $v_j = \Pi_{l=i_1}^{i_{m_T}} (v_j^{(l)})^{\alpha_j^{(l)}}$. One can verify that $u_j = u^{k_j} = g^{rk_j}$ and $v_j = v^{k_j} = H(w)^{k_j} h^{rk_j}$. Let $c_j = (u_j, v_j)$ and $m_{TG} = (\sigma_U(H(m_U)), \widetilde{\Delta}, \{c_j\}_{j=1}^{\kappa})$. TG then generates a signature σ_{GT} on the message m_{TG} (this task is trivial under the standard PKI assumption) and then sends (m_{TG}, σ_{TG}) to the $DBMS$.

- Upon receiving (m_{TG}, σ_{TG}), $DBMS$ checks the validity of the received message. If it is invalid, then terminates the protocol; otherwise, it decrypts $\widetilde{\Delta}$ to get Δ and broadcasts $\{u_j\}_{j=1}^{\kappa}$ to all token extractors within Δ via a secure multi-cast channel (such a multi-cast channel can be efficiently implemented in the context of group communication protocol).

  Each token extraction processor TEP_l computes $\widetilde{u}_j^{(l)} = u_j^{x_l}$ and sends $(TEP_j^{(l)}, \widetilde{u}_j^{(l)})$ to TE, where $TEP_j^{(l)}$ stands for the jth input processed by the lth token extraction processor. The computing results $\{TEP_j^{(l)}, \widetilde{u}_j^{(l)}\}_{l=i_1}^{i_{m_E}}$ are then sent back to the $DBMS$.

- Upon receiving $\{TEP_j^{(l)}, \widetilde{u}_j^{(l)}\}_{l=i_1}^{i_{m_E}}$ from all token extraction processors, the $DBMS$ computes $\widehat{u}_j = \Pi_{l=1}^{m_E} \widetilde{u}_j$. One can verify that $\widehat{u}_j = h^{rk_i}$. As a result, given $\widehat{u}_i$ and $\{v_j\}_{j=1}^{m_E}$, $DBMS$ is able to extract the search token $H(w)^{k_i}$. Let $\mathcal{D}$ be a set of encrypted documents so that $c(w) \in D$ for each $D \in \mathcal{D}$. $DBMS$ then sends $\mathcal{D}$ to the user U;

The data extraction phase

– Upon receiving $\mathcal{D}$, the user performs a decryption of message via the Shamir's threshold decryption protocol to obtain sk_i. Once obtains sk_i, the user U can decrypt the received ciphertexts.

This ends the description of our protocol.

3.2   The Proof of Security

The correctness of the protocol can be verified step by step and hence omitted. The rest of this section is to provide a proof of security defined in Sect. 3.

Lemma 1. *Let Δ be an arbitrary subset of $\{TEP_1, \cdots, TEP_m\}$, suppose at least one of the selected token extraction processor is honest then the proposed scheme is keyword hiding assuming that the decisional Diffie-Hellman problem is hard.*

Proof. $\mathcal{C}$ invokes n token extraction processors each of which takes as input **params** to output n pairs of public and secret keys (pk_{TEP_i}, sk_{TEP_i}) $(i = 1, \cdots, n)$. $\mathcal{C}$ then provides pk_{TEP_i} $(i = 1, \cdots, n)$ to $\mathcal{A}$. Let Δ be an arbitrary subset of $\{1, \cdots, n\}$ containing m public key indexes. Without loss of the generality, we assume that $\Delta = \{(g, pk_{TEP_1}), \ldots, (g, pk_{TEP_m})\}$, where $pk_{TEP_i} = g^{x_{TEP_i}}$. The challenger is allowed to corrupt $m - 1$ token extraction processors and obtains the corresponding secret keys x_{TEP_i} for $i = 1, \ldots, m - 1$. The challenger's target is to break the mth instance of the ElGamal encryption scheme.

Let h $(= pk_{TEP}^{(\Delta)}) = pk_{TEP_1} \times \cdots \times pk_{TEP_m}$. Let w_0 and w_1 be two keywords output by the adversary which is also known to the challenger. Let $c \leftarrow (g^r, H(w_b) \times h^r)$ (generated by the semantic security game of the underlying encryption scheme (g, pk_{TEP_m})). The challenger then forwards (w_0, w_1) and c to the adversary $\mathcal{A}$. The adversary outputs a guess $b' \in \{0, 1\}$. The challenger outputs what the adversary outputs.

Given $c = (g^r, H(w_b)h^r)$ and x_{TEP_i} for $i = 1, \ldots, m - 1$, the challenger is able to compute $(g^r, H(w_b)h_m^r)$ where $h_m = pk_{TEP_m}$. Hence if $Pr[b = b'] - 1/2|$ is a non-negligible then the mth instance of the underlying ElGamal encryption scheme is not semantically secure which contradicts the decisional Diffie-Hellman assumption.

Lemma 2. *Let $\Delta = \{TGP_1, \cdots, TGP_m\}$ be an arbitrary subset of total token generation processors. Suppose at least one of the selected token extraction processor is honest then the proposed scheme is token hiding assuming that the decisional Diffie-Hellman problem is hard and the underlying BH pseudorandom number generator is cryptographically strong.*

Proof. Let w be a keyword selected by $\mathcal{A}$. Let k_j be the key used to generate the search token at session $sid = j$. Assuming that up to $(m - 1)$ token generation processors are corrupted and the adversary obtains the corresponding shares, say $(k_j^{(1)}, \ldots, k_j^{(m-1)})$. The mth token generation processor remains honest at the session $sid = j$. Let $k_j^{(l)}$ be a secret share of k_j by TGP_l for $l = 1, \cdots, n_G$. Notice that the only knowledge applied to $H(w)$ is $k_j^{(l)}$. For fixed set of shares $(k_j^{(1)}, \ldots, k_j^{(m-1)})$, there is a one to one mapping (the Lagrange interpolation formula) between $k_j^{(l)}$ and k_j. As a result, $H(w)^{k_j}$ is random from the point view of the adversary if the underlying pseudo-random number generator is secure. As a result, the only information leaked is the computation of $H(w)^{k_j^{(l)}}$. Since G is a cyclic group, it follows that $H(w) = g^{r_w}$ for some $r_w \in [0, q - 1]$. Thus, $(g, H(w), g^{k_j^{(l)}}, H(w)^{k_j^{(l)}})$ is a Diffie-Hellman quadruple that is indistinguishable from the random quadruple. As a result, the advantage $Pr[b = b'] - 1/2|$ that the adversary outputs a correct guess is at most a negligible amount.

Lemma 3. *Let $\Delta = \{DEP_1, \cdots, DEP_m\}$ be an arbitrary subset of total data extraction processors. Suppose at least one of the selected data extraction processor is honest then the proposed scheme is data hiding assuming that the decisional Diffie-Hellman problem is hard.*

Proof. $\mathcal{C}$ invokes the data owner O which takes as input the system parameters **params** to output a pair of public and secret keys (pk_O, sk_O). The adversary $\mathcal{A}$ is given pk_O; At the session i, O takes **params** as input and generates a mask key K_i such that $K_i = t_i^{(1)}$, where the auxiliary string $t_i^{(1)}$ is the first part of the output t_i generated by the BH pseudo-random generator.

$\mathcal{C}$ invokes n data extraction processors each of which takes as input **params** to output n pairs of public and secret keys (pk_{DEP_i}, sk_{DEP_i}) $(i = 1, \cdots, n)$. $\mathcal{C}$ then provides pk_{DEP_i} to $\mathcal{A}$. Let $K_i^{(l)}$ be the share of $t_i^{(1)}$ via the Lagrange interpolation formula for $l = 1, \cdots, n$. Let Δ be an arbitrary subset of $\{DEP_1, \cdots, DEP_n\}$ and m_0 and m_1 be two documents all selected by $\mathcal{A}$. We assume that the adversary can corrupt up to $(m - 1)$ data extraction processors and obtains the corresponding secret shares $K_i^{(l)}$ for $l = 1, \ldots, m - 1$. The data m_b is then encrypted under K_i, i.e., $c_i = (u_i, v_i)$, where $u_i = g^r$ and $v_i = m_b K_i^r$ (here for simplicity, we assume that m_b is encrypted under K_i directly). The adversary is given c_i. The adversary obtains the $(m - 1)$ secret shares each of which is holden by the corrupted parties say $DEP_1, \ldots, DEP_{m-1}$. Notice that $K_i^r = u_i^{K_i^{(1)} \alpha_1} \times \cdots \times u_i^{K_i^{(m-1)} \alpha_{m-1}} \times u_i^{K_i^{(m)} \alpha_m}$, where α_i is the ith coefficient of the Lagrange Interpolation formula. Thus, $m_b K_i^r$ is a random value from the point view of the adversary. As a result, the advantage $Pr[b = b'] - 1/2|$ that the adversary outputs a correct guess is at most a negligible amount.

Based on the lemmas above and we claim the following main result

Theorem 1. *The proposed multi-key searchable encryption is semantically secure under the joint assumptions that the decisional Diffie-Hellman problem is hard in Z_p^* and the underlying pseudo-random number generator deployed is cryptographically strong.*

4   Conclusion

In this paper, an efficient multi-user, multi-key searchable encryption scheme is presented and analyzed. Our design is simple, scalable, adaptable and sustainable. The processors are distributed to provide high reliability with limited storage, communication and computation overhead by the threshold cryptographic system.

References

1. Bao, F., Deng, R.H., Ding, X., Yang, Y.: Private query on encrypted data in multi-user settings. In: Chen, L., Mu, Y., Susilo, W. (eds.) ISPEC 2008. LNCS, vol. 4991, pp. 71–85. Springer, Heidelberg (2008)
2. Barak, B., Halevi, S.: A model and architecture for pseudo-random generation with applications to dev random. In: ACM Conference on Computer and Communications Security, pp. 203–212 (2005)
3. Bellare, M., Boldyreva, A., O'Neill, A.: Deterministic and efficiently searchable encryption. In: Menezes, A. (ed.) CRYPTO 2007. LNCS, vol. 4622, pp. 535–552. Springer, Heidelberg (2007)
4. Bethencourt, J., Song, D.X., Waters, B.: New techniques for private stream searching. ACM Trans. Inf. Syst. Secur. **12**(3), 16 (2009)
5. Boneh, D., Di Crescenzo, G., Ostrovsky, R., Persiano, G.: Public key encryption with keyword search. In: Cachin, C., Camenisch, J.L. (eds.) EUROCRYPT 2004. LNCS, vol. 3027, pp. 506–522. Springer, Heidelberg (2004)
6. Boneh, D., Franklin, M.: Identity-based encryption from the weil pairing. In: Kilian, J. (ed.) CRYPTO 2001. LNCS, vol. 2139, pp. 213–229. Springer, Heidelberg (2001)
7. Boneh, D., Waters, B.: Conjunctive, subset, and range queries on encrypted data. In: Vadhan, S.P. (ed.) TCC 2007. LNCS, vol. 4392, pp. 535–554. Springer, Heidelberg (2007)
8. Chor, B., Kushilevitz, E., Goldreich, O., Sudan, M.: Private information retrieval. J. ACM **45**(6), 965–981 (1998)
9. Cao, N., Wang, C., Li, M., et al.: Privacy-preserving multi-keyword ranked search over encrypted cloud data. IEEE Trans. Parallel Distrib. Syst. **25**(1), 222–233 (2014)
10. Curtmola, R., Garay, J.A., Kamara, S.: Ostrovsky, R.: Searchable symmetric encryption: improved definitions and efficient constructions. In: ACM Conference on Computer and Communications Security, pp. 79–88 (2006)
11. Gertner, Y., Ishai, Y., Kushilevitz, E., Malkin, T.: Protecting data privacy in private information retrieval schemes. J. Comput. Syst. Sci. **60**(3), 592–629 (2000)
12. Gathegi, J.N.: Clouding big data: information privacy considerations. In: Gathegi, J.N., Tonta, Y., Kurbanoğlu, S., Al, U., Taşkin, Z. (eds.) Challenges of Information Management Beyond the Cloud. Springer, Heidelberg (2014)
13. Goh, E.-J.: Secure indexes. IACR Cryptology ePrint Archive, p. 216 (2003)
14. Hoffstein, J., Pipher, J., Silverman, J.H.: NTRU: a ring-based public key cryptosystem. In: Buhler, J.P. (ed.) ANTS 1998. LNCS, vol. 1423, pp. 267–288. Springer, Heidelberg (1998)
15. Kamara, S., Papamanthou, C., Roeder, T.: Dynamic searchable symmetric encryption. In: ACM Conference on Computer and Communications Security, pp. 965–976 (2012)

16. Hahn, F., Kerschbaum, F.: Searchable encryption with secure, efficient updates. In: Proceedings of the 2014 ACM SIGSAC Conference on Computer, Communications Security. ACM, pp. 310–320 (2014)

17. López-Alt, A., Tromer, E., Vaikuntanathan, V.: On-the-fly multiparty computation on the cloud via multikey fully homomorphic encryption. In: STOC, pp. 1219–1234 (2012)

18. Liu, J.K., Au, M.H., Huang, X., Susilo, W., Zhou, J., Yu, Y.: New insight to preserve online survey accuracy and privacy in big data era. In: Kutyłowski, M., Vaidya, J. (eds.) ICAIS 2014, Part II. LNCS, vol. 8713, pp. 182–199. Springer, Heidelberg (2014)

19. Malkin, T.: Secure computation for big data. In: Sahai, A. (ed.) TCC 2013. LNCS, vol. 7785, pp. 355–355. Springer, Heidelberg (2013)

20. Ostrovsky, R., Skeith, W.E.: Private searching on streaming data. J. Cryptology 20(4), 397–430 (2007)

21. Pappas, V., Raykova, M., Vo, B., Bellovin, S.M., Malkin, T.: Private search in the real world. In: ACSAC, pp. 83–92 (2011)

22. Popa, R.A., Zeldovich, N.: Multi-key searchable encryption. IACR Cryptology ePrint Archive, p. 508 (2013)

23. Popa, R., Stark, E., Helfer, J., Valdez, S., Zeldovich, N., Kaashoek, M.F., Balakrishnan, H.: Building web applications on top of encrypted data using mylar. In: NSDI (USENIX Symposium of Networked Systems Design and Implementation) (2014)

24. Orencik, C., Selcuk, A., Savas, E., et al.: Multi-Keyword search over encrypted data with scoring, search pattern obfuscation. Int. J. Inf. Secur. 1–19 (2015)

25. Rabin, M.O.: How to exchange secrets by oblivious transfer. Technical report TR-81, Aiken Computation Laboratory, Harvard University (1981)

26. Raykova, M., Cui, A., Vo, B., Liu, B., Malkin, T., Bellovin, S.M., Stolfo, S.J.: Usable, secure, private search. IEEE Secur. Privacy 10(5), 53–60 (2012)

27. Raykova, M., Vo, B., Bellovin, S.M., Malkin, T.: Secure anonymous database search. In: CCSW, pp. 115–126 (2009)

28. Samanthula, B.K., Elmehdwi, Y., Howser, G., Madria, S.: A secure data sharing and query processing framework via federation of cloud computing. Inf. Syst. 48, 196–212 (2015)

29. Song, D.X., Wagner, D., Perrig, A.: Practical techniques for searches on encrypted data. In: IEEE Symposium on Security and Privacy, pp. 44–55 (2000)

30. Tang, Y., Liu, L.: Privacy-preserving multi-keyword search in information networks (2015)

31. Yang, Y.: Towards multi-user private keyword search for cloud computing. In: IEEE CLOUD, pp. 758–759 (2011)

32. Yang, J.J., Li, J.Q., Niu, Y.: A hybrid solution for privacy preserving medical data sharing in the cloud environment. Future Gener. Comput. Syst. 43, 74–86 (2015)

Another Look at Aggregate Signatures: Their Capability and Security on Network Graphs

Naoto Yanai[1,2]([✉]), Masahiro Mambo[3], Kazuma Tanaka[4],
Takashi Nishide[4], and Eiji Okamoto[4]

[1] Osaka University, Osaka, Japan
yanai@ist.osaka-u.ac.jp
[2] AIST, Osaka, Japan
[3] Kanazawa University, Kanazawa, Japan
[4] University of Tsukuba, Tsukuba, Japan
tanaka@cipher.risk.tsukuba.ac.jp, {nishide,okamoto}@risk.tsukuba.ac.jp

Abstract. Aggregate signatures are digital signatures where n signers sign n individual documents and can aggregate individual signatures into a single short signature. Although aggregate signatures are expected to enhance the security of network applications, the capability and the security of aggregate signatures have not yet been discussed when the signatures are generated by a group of signers whose relationships are expressed as network. In this paper, we take into account the fact that various network applications can be mathematically idealized as network called network graphs, and discuss the properties of aggregate signatures on network graphs. We show that it is difficult to apply aggregate signatures to the network graphs. More precisely, we show that sequential aggregate signatures (Eurocrypt 2004) are incompatible with the network graphs and also general aggregate signatures (Crypto 2003) are broken by some generic attack. Additionally, we propose two generic approaches to overcoming the problems: restricting the number of signers and utilizing ring homomorphism, and give a security proof of aggregate signatures in each of these approaches.

Keywords: Aggregate signatures · Sequential aggregate signatures · Cryptographic protocols · Provable security · Graph theory

1 Introduction

Motivation. An aggregate signature scheme [7] is a cryptographic primitive where each signer signs an individual data and these individual signatures can be aggregated into a single short signature. The primitive has been expected to provide many applications such as enhancement of security of routing protocols in networks, e.g., border gateway protocol (BGP) [29]. In particular, some secure routing protocols need to sign data updated in each device, and the aggregate signature scheme has been considered as a suitable solution for compressing a memory storage and a communication overhead.

© Springer International Publishing Switzerland 2016
M. Yung et al. (Eds.): INTRUST 2015, LNCS 9565, pp. 32–48, 2016.
DOI: 10.1007/978-3-319-31550-8_3

Nevertheless, when aggregate signatures are used in network applications of a network system, its security has not been evaluated correctly in all cases as shown in this paper. In general, behavior of cryptographic protocols in applications should be discussed rigorously for constructing a secure system. Otherwise, even with the use of provably secure cryptographic protocol, its system may be broken by manipulating some part of applications in the system. In fact, Sun et al. [32] showed several attacks against the Tor networks [11] by controlling BGP. Tor is an anonymity system to protect user identities from both untrusted destinations and third parties in the Internet and it utilizes SSL/TLS as the privacy mechanism [32]. Their attacks are launched by operating autonomous systems (ASes) maliciously and can compromise privacy of users. Although the security of TLS protocols has been proven in several papers [15,22], the attack is successfully conducted by manipulating BGP operation.

In the similar vein, discussing behavior of aggregate signatures on networks is crucial for constructing secure network systems. In our discussion, we adopt mathematically idealized networks called *network graphs*, and discuss properties on the network graphs. Current network applications can be abstracted as network graphs, and thus we expect that discussing aggregate signatures on the network graphs are useful for analyzing the capability and the security of aggregate signatures in the existing network applications.

Our Contribution. We show two negative results for utilizing general aggregate signatures and sequential aggregate signatures on network graphs, and two potential approaches to overcoming these results.

The first negative result is that a sequential aggregate signature scheme [27] is unsuitable in network graphs. A sequential aggregate signature scheme is more efficient than a general aggregate signature scheme [7] since broadcast of signatures is unnecessary for combining them. We identity a case where such a sequential aggregate signature scheme does not work well. The second negative result is insecurity of a general aggregate signature scheme [7]. We consider our attack is crucial for many network environments. In particular, an adversary can forge signatures on any message by generating some malicious network graph. This insecurity implies that applications such as aggregated-path authentication protocol [37] are insecure. Note that the security proof is correct, and our attack is out of their models. Meanwhile, we consider our attack is sufficiently executable for many network environments.

We then propose two generic schemes which converts any general aggregate signature scheme to one overcoming the insecurity problem. In particular, our approaches are to (1) restrict the number of signers and (2) utilize a ring homomorphism. The former approach is to prevent an adversary from generating the malicious network graph without decreasing the efficiency and is for a relatively small-scale operation. In a large-scale network, the second approach can be utilized. It allows us to overcome the problem without the restriction in the number of signers. We also note that the second approach is secure, but constructing an instantiation of a ring homomorphism is still an open problem.

Potential Application. A potential application of this work is secure routing protocols with multi-path setting. In recent routing protocols, several secure routing protocols [19,21,25] have been proposed, which guarantee the validity of routing information. Here, the multi-path setting is a network structure containing multiple paths from any source to its destination. The multi-path setting provides many positive results in routing protocols according to Valere [34].

Related Work. Aggregate signatures are classified into two types, i.e., general aggregate signatures [7] and sequential aggregate signatures [27]. General aggregate signatures need a general aggregation algorithm where anyone can compress individual signatures, and require each signer to broadcast individual signatures to its co-signers. General aggregate signatures can be generically constructed via full domain hash signatures, and there are the BGLS03 scheme [7] and the BNN07 scheme [3] in the random oracle model. Meanwhile, in the standard model, there are the AGH10 scheme [1] with stateful setting, and the RS09 scheme [30] and the HSW13 scheme [17] with stateless setting by multilinear maps. As a more recent result, a general aggregate signature scheme in the standard model can be obtained via any full-domain hash construction and indistinguishability obfuscations [18]. Moreover, universal signature aggregator [16] where any signatures can be aggregated has been proposed.

Sequential aggregate signatures utilize an aggregate-signing algorithm where signature generation is executed with signature aggregation. According to Lysyanskaya et al. [27], sequential aggregate signatures can be generically constructed via trapdoor permutation chains. That is, only signers owning secret keys can compress signatures. The LMRS04 scheme [27] is the first and generic scheme in the random oracle model, and then the LOSSW06 scheme [26] with the Waters hash function [35] has been proposed. Sequential aggregate signatures seem to be excellent with a construction without the random oracles, and there are the Schröder09 scheme [31] and the LLY13-2 scheme [23] with the CL signatures [10], and the LLY13-1 scheme [24] with the dual system methodology [36]. Meanwhile, constructions with the random oracles are aimed to provide more serviceable operation. For instance, there are the Neven08 scheme [28] with message-recovery, the FSL12 scheme [12] with history-freeness, the BGR12 scheme [8] with lazy verifications, and the BGOY07/10 scheme [5,6] and the GLOW12 scheme [14] with ID-based setting. The BGOY07 scheme was broken in [20] and then the security was fixed in the BGOY10 scheme based on a new interactive assumption. The GLOW12 version is an improved scheme of the BGOY10 scheme whose security is reduced to a static assumption by the dual system methodology.

2   Aggregate Signatures

We recall algorithms of aggregate signatures and their security. First, we describe notations in this paper. We denote by (sk_i, pk_i) a pair of a secret key and a public key of the ith signer. Here, we assume that a single pair of keys is corresponding to a single signer and represent the signer as pk_i for all i. We also denote by $\mathcal{M}$ a message space, by $m_i \in \{0,1\}^*$ a message to be signed by pk_i, by σ_i a signature generated by pk_i, and by $\perp$ an error.

2.1   General Aggregate Signatures

Syntax. A general aggregate signature scheme is defined as follows:

Setup Given a security parameter 1^k, output a public parameter $para$.
KeyGen Given $para$, output a secret key sk_i and a public key pk_i.
Sign Given $para$, sk_i and a message $m_i \in \mathcal{M}$, generate a signature σ_i and then output σ_i. If any error happens, then output an error $\perp$.
Aggregate Given $para$ and i-tuples $(\{(pk_j, m_j, \sigma_j)\}_{j=1}^i)$ of public keys, messages and signatures, output a single short signature σ.
Verify Given $(para, \{(pk_j, m_j)\}_{j=1}^i, \sigma)$ as input, output $\top$ or $\perp$.

Definition 1 (Correctness). The correctness of the general aggregate signature scheme is defined as follows: for all k, all $para \leftarrow$ **Setup**(1^k), all $(sk_i, pk_i) \leftarrow$ **KeyGen**$(para)$ and all integer i, the following equation holds if no $\perp$ is returned by **Sign**$(para, sk_j, m_j)$ for any $j \in [1, i]$:

$$\top = \textbf{Verify} \left(\begin{array}{c} para, \{(pk_j, m_j)\}_{j=1}^i, \\ \textbf{Aggregate}\left(para, \{(pk_j, m_j, \textbf{Sign}(para, sk_j, m_j))\}_{j=1}^i\right) \end{array} \right).$$

If there exists **Sign**$(para, sk_j, m_j)$ whose output is $\perp$, output of the above equation is $\perp$.

Definition of Security. We recall the security of general aggregate signatures. In this model, entities are a challenger $\mathcal{C}$ and an adversary $\mathcal{A}$ with a security parameter 1^k as input. $\mathcal{C}$ has a public key list $\mathcal{L}$. Here, we denote by $x^{(h)}$ h-th query for all x in the following game:

Setup $\mathcal{C}$ runs the setup algorithm to obtain $para$ and the key generation algorithm to obtain a pair (sk^*, pk^*) of a challenge key. Then $\mathcal{C}$ interacts with $\mathcal{A}$ with $(para, pk^*)$ as follows:
Key Certification Query Given a pair $(pk_i^{(h)}, sk_i^{(h)})$ by $\mathcal{A}$, $\mathcal{C}$ registers $pk_i^{(h)}$ in $\mathcal{L}$ if they are a valid pair. If not, output an error symbol $\perp$.
Signing Query Given a message $m^{(h)}$, $\mathcal{C}$ generates a signature σ on $m^{(h)}$ by the signing algorithm. Then, $\mathcal{C}$ returns σ.
Output Given a forgery $(\{(m_j^*, pk_j^*)\}_{j=1}^i, \sigma^*)$ by $\mathcal{A}$, $\mathcal{C}$ checks if the following conditions hold; the **Verify**$(para, \{(pk_j^*, m_j^*)\}_{j=1}^i, \sigma_i^*)$ outputs $\top$; for some $j \in [1, i]$, pk_j is exactly corresponding to pk^*; each public key in $\{pk_j^*\}_{j=1}^i$ is registered in $\mathcal{L}$ except for pk^*; for some i corresponding to pk^*, $\mathcal{A}$ did not query m_i^* to the signing oracle. $\mathcal{A}$ wins if all of these conditions hold.

Definition 2. We say that an adversary $\mathcal{A}$ breaks a general aggregate signature scheme with $(t, q_c, q_s, q_h, n, \epsilon)$ if $\mathcal{A}$ who does not know sk^* can win the game described above with a success probability greater than ϵ within an execution

time t. Here, $\mathcal{A}$ can access to the signing oracle at most q_s times, the key certification oracle at most q_c times and random oracles at most q_h times, and n is
the number of signers included in the forgery. We say that a general aggregate
signature scheme is $(t, q_c, q_s, q_h, n, \epsilon)$-secure if there is no adversary who breaks
the scheme with $(t, q_c, q_s, q_h, n, \epsilon)$.

2.2 Sequential Aggregate Signatures

Many parts of a syntax and security of sequential aggregate signatures are the
same as those of general aggregate signatures. Therefore, we describe different
parts from the previous section, and a full syntax and a definition of the security
are omitted due to the space limitation.

Syntax. A sequential aggregate signature scheme consists of four algorithms,
Setup, **KeyGen**, **Aggregate-Sign** and **Verify**. These are exactly the same
as that of general aggregate signatures except for **Aggregate-Sign** algorithm.
Therefore, we describe only **Aggregate-Sign** algorithm.

Aggregate-Sign Given $para, sk_i$, a message $m_i \in \mathcal{M}$, $i-1$ pairs $\{(pk_j, m_j)\}_{j=1}^{i-1}$
of public keys and messages for previous signers, and their signature σ_{i-1},
generate an aggregate signature σ_i and output σ_i. If any input is an error,
then output an error symbol $\perp$.

Definition 3 (Correctness). The correctness of the sequential aggregate signature scheme is defined as follows: for all security parameter k, all $para \leftarrow$
Setup(1^k), all $(sk_i, pk_i) \leftarrow$ **KeyGen**$(para)$ and all integer i, the following equation holds if verification of a given signature σ_{i-1} is $\top$:

$$\top = \textbf{Verify} \left(para, \{(pk_j, m_j)\}_{j=1}^{i}, \textbf{Aggregate-Sign} \left(\begin{array}{c} para, sk_i, m_i, \\ \{(pk_j, m_j)\}_{j=1}^{i-1}, \sigma_{i-1} \end{array} \right) \right).$$

If verification of a given σ_{i-1} is $\perp$, then **Aggregate-Sign** outputs $\perp$.

3 Network Graph

In this section we define network graphs. Network graphs are defined as an
extension of series-parallel graphs [4], and we first recall the definition of series-
parallel graphs [4].

3.1 Series-Parallel Graph

Definition of Series-Parallel Graph. Let $\mathcal{G}$ be a set of graphs. A series-
parallel graph is a graph generated by recursively applying either a serial graph
or a parallel graph in an arbitrary order. More specifically, a series-parallel graph
$G(I, T)$, which starts at the initial vertex I and terminates at the terminal vertex
T, is defined as follows:

$G(I, T)$ is generated either by the following step 1 or step 2.

1. With a unique label i in $\mathcal{G}$, $G_i(I_i, T_i)$ is composed of one edge connecting I_i and T_i. We call such a graph an atomic graph and denote it by $\phi_i \in \mathcal{G}$.
2. For the step 2, either the following step (a) or step (b) is executed.
   (a) (Parallel Graph) Given n graphs $G_i(I_i, T_i)$ for $1 \leq i \leq n$, construct $G(I, T)$ by setting $I = I_1 = I_2 = \cdots = I_n$ and $T = T_1 = T_2 = \cdots = T_n$.
   (b) (Serial Graph) Given n graphs $G_i(I_i, T_i)$ for $1 \leq i \leq n$, construct $G(I, T)$ by setting $I = I_1, T_1 = I_2, \cdots, T_{n-1} = I_n$ and $T_n = T$.
   Intuitively, in the above definitions, constructing $G(I, T)$ means compositions of n atomic graphs $\phi_i \in \mathcal{G}$ for $i = [1, n]$ either as a serial one or a parallel one.

Composition of Graphs. For two graphs $\phi_1, \phi_2 \in \mathcal{G}$, we define a composition of parallel graphs as $\phi_1 \cup \phi_2$ and the composition of serial graphs as $\phi_1 \cap \phi_2$. In other words, $\phi_1 \cup \phi_2$ means to construct $G(I, T)$ by setting $I = I_1 = I_2$ and $T = T_1 = I_2$, and $\phi_1 \cap \phi_2$ means to construct $G(I, T)$ by setting $I = I_1, T_1 = I_2$ and $T_2 = T$. We denote by $\mathcal{T}(i)$ a set of graphs connecting to the initial vertex I_i of ith graph in such a way that $\mathcal{T}(i) = \{x \mid I_i = T_x \land 1 \leq x < i \land G_x(I_x, T_x) \subset \psi_n\}$, by $\mathcal{I}(i)$ a set of graphs connecting to the terminal vertex T_i of ith graph in such a way that $\mathcal{I}(i) = \{x \mid T_i = I_x \land i < x \leq n \land G_x(I_x, T_x) \subset \psi_n\}$, by $\{a_j\}_{j \in \mathcal{T}(i)}$, for all a, all a_j for $j \in \mathcal{T}(i)$. In other words, for all graphs, the composition of a graph ϕ_i and ψ_j for $j \in \mathcal{T}(i)$ can be denoted by $\psi_i := \phi_i \cap \left(\bigcup_{j \in \mathcal{T}(i)} \psi_j \right) = \phi_i \cap \psi_{\mathcal{T}(i)}$ where $\bigcup_x$ means iterations of the operation $\cup$ for all x. Similarly, $\bigcap_x$ can be defined as iteration of $\cap$.

Weight of Graph. We define a weight function $\omega_i(\psi_n)$ that represents a weight of each label i for a graph ψ_n for all i, n. Intuitively, $\omega_i(\psi_n)$ means the number of paths including an edge with a label i from I_i to T_n for ψ_n.

3.2   Definition of Network Graph

We define a network graph by extending a series-parallel graph. That is, a network graph inherits all properties of a series-parallel graph. Here, an edge of a series-parallel graph corresponds to a "network entity" such as an Internet service provider. For any network graph, each entity has a unique edge with a unique index representing the position of the entity, i.e., an index i in ψ_n corresponds to the ith entity in the whole network graph for all i, n. Hereafter, for all i, n, we denote by ψ_n a network graph consisting of n entities, by $\mathcal{T}(i)$ a set of entities connecting to the initial vertex I_i of the ith entity in such a way that $\mathcal{T}(i) = \{x | I_i = T_x \land 1 \leq x < i \land G_x(I_x, T_x) \subset \psi_n\}$, by $\mathcal{I}(i)$ a set of entities connecting to the terminal vertex T_i of ith entity in such a way that $\mathcal{I}(i) = \{x | T_i = I_x \land i < x \leq n \land G_x(I_x, T_x) \subset \psi_n\}$, by $\{a_j\}_{j \in \mathcal{T}(i)}$, for all a, all a_j for $j \in \mathcal{T}(i)$. We also define an operation $\dot{\subset}$ where $i \dot{\subset} \psi_n$ means extractions of indexes from ψ_n.

Instantiation of Network Graph. An instantiation of the network graphs described above is a routing protocol with multi-path setting where each router owns multiple paths to any destination. We briefly describe their intuitions below. Firstly, in the routing protocol, each network entity connects with multiple entities and a network graph has no loop structure in general. These properties are exactly corresponding to that of the network graph. Moreover, in a viewpoint of a source to send packets, route information to its destination corresponds to a series-parallel graph whose whole terminal edge is the destination. Namely, we consider many routing protocols with the multi-path setting can be represented as the network graph.

4   Do Aggregate Signatures Work on Network Graph?

We show negative results in utilizing aggregate signatures on network graphs. We first describe how to apply aggregate signatures to network graphs, and then show several theorems with respect to the negative results.

4.1   Application of Aggregate Signatures to Network Graph

Hereinafter, we suppose that a signer for aggregate signatures corresponds to an edge for network graphs for the sake of convenience. We also suppose that aggregate signatures are generated along with network graphs: more specifically, for any whole graph from a source to its destination, signatures are individually propagated on each path in the graph and are aggregated in any parallel part of the graph. Based on the definitions in Sect. 3, for any signer i, its co-signers for aggregating signatures are corresponding to a set $\mathcal{T}(i)$ of edges which connect to its initial vertex I_i. In other words, the ith signer generates a signature after he/she receives signatures from previous signers belonging to $\mathcal{T}(i)$, and then aggregates these signatures into a single signatures. These are natural operations along with network applications such as secure routing protocols. For example, each router in secure routing protocols signs routing information to guarantee the validity and aggregates the collected signatures in order to save the size of memory resource.

More specifically, when a signer receives signatures from co-signers along with $\mathcal{T}(i)$, he/she utilizes **Sign** and **Aggregate** algorithms in a general aggregate signature scheme while **Aggregate-Sign** algorithm in a sequential aggregate signature scheme, respectively. We also note that, if there are multiple signers at the end of a graph, an aggregate signature on the whole graph is an aggregation of a signature for each path. For the sake of convenience, we show the following figures, Figs. 1 and 2, as examples of the use of aggregate signatures on network graphs.

4.2   (In)security of General Aggregate Signatures

We describe the first negative result with some scheme. The following scheme is a generic construction of general aggregate signatures and the existing schemes are included in the construction.

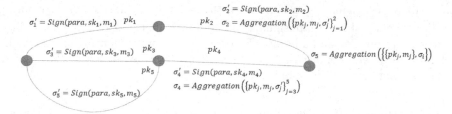

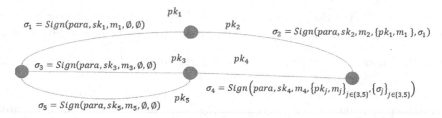

Fig. 1. How to use general aggregate signatures on network graph

Fig. 2. How to use sequential aggregate signatures on network graph

Generic Construction. A general aggregate signature scheme is generically constructed by utilizing a full-domain hash construction with bilinear maps. Here, the bilinear maps are defined as follows: Let $\mathbb{G}$ and $\mathbb{G}_T$ be groups with the same prime order p. A bilinear map $e : \mathbb{G} \times \mathbb{G} \to \mathbb{G}_T$ is a map such that the following conditions hold, where g is a generator of $\mathbb{G}$: (Bilinearity) For all $u, v \in \mathbb{G}$ and $a, b \in \mathbb{Z}_p^*$, $e(u^a, v^b) = e(u, v)^{ab}$; (Non-degeneracy) For any generator $g \in \mathbb{G}$, $e(g, g) \neq 1_{\mathbb{G}_T}$, $1_{\mathbb{G}_T}$ is an identity element over $\mathbb{G}_T$; (Computable) There is an efficient algorithm to compute $e(u, v)$ for any $u, v \in \mathbb{G}$. We say that $\mathbb{G}$ is a bilinear group if all these conditions hold, and we assume that the discrete logarithm problem (DLP) in bilinear groups is hard. We call the parameter $(p, \mathbb{G}, \mathbb{G}_T, e)$ *pairing parameter*.

Setup Given a security parameter 1^k, generate a generator $g \in \mathbb{G}$ and choose a full-domain hash function $H : \{0, 1\}^* \to \mathbb{G}$. Then, output g and H as a public parameter *para*.

KeyGen Given *para*, generate $x_i \in \mathbb{Z}_p^*$ as sk_i and compute $y_i = g^{x_i}$ as pk_i. Output sk_i and pk_i.

Sign Given *para*, sk_i and a message $m_i \in \{0, 1\}^*$, compute $h_i = H(m_i)$ and $\sigma_i = h_i^{x_i}$. Then output σ_i.

Aggregate Given *para* and i-tuples $(\{(pk_j, m_j, \sigma_j\}_{j=1}^i)$ of public keys, messages and signatures, output a single short signature $\sigma = \prod_{j=1}^i \sigma_j$.

Verify Given $(para, \{pk_j, m_j\}_{j=1}^i, \sigma)$ as input, output $\top$ if $e(\sigma, g) = \prod_{j=1}^i e(h_j, y_j)$ where $h_j = H(m_j)$ for all j. Otherwise, output $\bot$.

One might think that the above construction is specific and not generic. We note that the scheme is almost generic since a full-domain hash signature scheme gives rise to aggregate signatures according to Hohenberger et al. [16]. In fact, all of the existing general aggregate signature schemes [1,13,17,30] are constructed by replacing a public key and a family of hash functions with their instantiations in each scheme. Therefore, discussion on the scheme described above can be applied to the existing schemes in a similar manner.

Negative Result. We pointed out that utilizing general aggregate signatures on network graphs induces some problem in security. That is, malicious signers can forge signatures on any message by generating a malicious graph.

Theorem 1. There exists an adversary which breaks the generic construction of general aggregate signatures on a network graph with non-negligible probability.

Proof. We show some attack as a counterexample against the security. As described in the previous section, signatures for each signer are sent to all of the following paths on a network graph. In this case, the signatures are amplified by their aggregation step on the following paths since a number of signatures in proportion to the number of paths are sent individually. More specifically, let a target signature be Alice and her signature be σ_A. Then, the value of σ_A is amplified by a product computation on the aggregation algorithm because an aggregate signature on each parallel path includes σ_A. Since the number of paths from any source to its destination is obtained as the weight of the graph described in Sect. 3.1, the aggregate signatures are given as $\sigma = \prod_{j=1}^{i} \sigma_j^{\omega_j(\psi_i)}$ for all i. Here, for some j, σ_j is corresponding to σ_A, and we denote by $\omega_A(\psi_i)$ the weight function of Alice for the sake of convenience. If the weight function $\omega_A(\psi_i)$ of Alice's signature σ_A is equal to zero under the modulo p, the value of $\sigma_A^{\omega_j(\psi_i)}$ is equal to 1 on any message since p is the order of a group $\mathbb{G}$. That is, an adversary can forge a signature on any message by outputting 1 via generating a malicious graph. The success probability of the adversary for generating such a graph is non-negligible. In order to guarantee the statement, we recall some lemma for the weight of a series-parallel graph. Since network graphs inherit properties of a series-parallel graph as described in Sect. 3, the lemma is applicable to network graphs.

Lemma 1 ([33], Lemma 2.3). *For any $i, n \in \mathbb{Z}$ and any graph $\psi_i \in \mathcal{G}$, it holds $\omega_i(\psi_n) \leq 3^{\#\psi_n/3}$.*

For the weight of a series-parallel graph, the lemma gives its upper bound with a fixed number of edges. In particular, the upper bound of the weight function is $3^{\#\psi_i/3}$ where $\#\psi_i$ is the number of edges, i.e., signers, for any graph ψ_i, and the attack is successful if $3^{\#\psi_i/3} = p$. For instance, p is 160-bit prime for 80-bit security and the adversary is always able to execute the attack by colluding with 300 signers.   □

We note that security proofs of the existing works are correct even against our theorem. The attack described above is executable because it is done on network graphs, and is out of the scope of their works.

4.3   (Un)availability of Sequential Aggregate Signatures

Generic Construction of Sequential Aggregate Signatures. A sequential aggregate signature scheme can be generally constructed from any trapdoor (permutation) chains: more precisely, **Aggregate-Signing** algorithm takes a trapdoor as a secret key and a given aggregate signature as input, and outputs a new aggregate signature by the trapdoor. Here, we denote by $\odot$ a group operation for the permutations:

Setup Given a security parameter 1^k, choose a full-domain hash function $H :$ $\{0,1\}^* \to D$ where D is a domain of any permutation. output H as a public parameter $para$.

KeyGen Given $para$, generate a trapdoor permutation π_i and its inverse permutation π_i^{-1}. output π_i^{-1} as sk_i and π_i as pk_i.

Aggregate-Sign Given $para, sk_i$, a message $m_i \in \{0,1\}^*$, $i-1$ pairs $\{(pk_j, m_j)\}_{j=1}^{i-1}$ of public keys and messages for previous signers, and their signature σ_{i-1}, compute $h_i = H(pk_1 \| \cdots \| pk_i, m_1 \| \cdots \| m_i)$ and generate an aggregate signature $\sigma_i = \pi_i^{-1}(h_i \odot \sigma_{i-1})$ where set $\sigma_0 = 1$ for $i = 1$. Output σ_i.

Verify Given $(para, \{pk_j, m_j\}_{j=1}^i, \sigma_i)$ as input, compute for all j $h_j = H(pk_1 \| \cdots \| pk_i, m_1 \| \cdots \| m_i)$ and then $\sigma_{j-1} = \pi_j(\sigma_j) \odot h_j^{-1}$. If $\sigma_0 = 1$, output $\top$. Otherwise, output $\bot$.

We note that the construction described above is generic in the sense that a scheme is constructed via any trapdoor chains (π_i, π_i^{-1}). In other words, we can take into account only constructing a new variant of (π_i, π_i^{-1}) in order to propose a sequential aggregate signature scheme. In a fact, an RSA-based construction [27] is a simple instantiation and bilinear-based constructions [26,31] can be found as kinds of the generic construction by replacing π_i and h_i with bilinear maps and map-to-point functions. Although our negative result is applicable to only schemes based on the generic construction, all of the existing constructions have been on this framework.

Negative Result. When we utilize a sequential aggregate signature scheme on network graphs, we have the following theorem.

Theorem 2. The correctness of the generic construction of sequential aggregate signatures does not hold on a network graph with a non-negligible probability if there exists a parallel path and the size of signatures is fixed with respect to the number of parallel paths.

Proof. We show some counterexamples against the correctness of the scheme described above. The intuition of the proof is that the aggregate-signing algorithm is constructed to combine signatures by a trapdoor, i.e., a secret key, of a signer and verify the validity by permutations of a public key. That is, in any parallel graph, while an individual signature is generated for each path, these signatures cannot be aggregated for a terminal signer of the parallel graph.

More specifically, we describe the statement with Fig. 2. In that figure, co-signers of David are Charlie and Edwin and David receives individual signatures from them. Then, David cannot aggregate their signatures. If David forcibly tries to generate a new aggregate signature by combining the Charlie's aggregate signature σ_3 and the Edwin's aggregate signature σ_5, then David's aggregate signature σ_4 is computed as $\sigma_4 = \sigma_3 \odot \sigma_5 \odot h_4$ where h_4 is a hashed digest of David. In this case, a verifier cannot verify the validity of σ_4 since he/she cannot compute inverse permutations of signature chains on the verification algorithm: more precisely, the verification algorithm requires a verifier to extract σ_3 and σ_5 from σ_4 via inverse permutations, but the verifier cannot separate σ_3 and σ_5 from $\sigma_3 \odot \sigma_5$ since he/she does not know σ_3 and σ_5. The success probability of the verification in such a situation is obtained as that of a correct separation between $\sigma_3 \odot \sigma_5$, and is depending on size of the range of each permutation π_i, i.e., $\frac{1}{|\pi_i|}$. Thus, the verification fails with an overwhelming probability. In order to recover the incorrectness of the generic construction, σ_3 and σ_5 are required to be kept separate. That is, in a whole graph, a large number of signatures in proportion to the number of parallel paths is necessary. Another counterexample is on the end of the whole graph in Fig. 2. Since sequential aggregate signatures do not contain an aggregation algorithm individually, signatures on the end of the graph cannot be aggregated. Namely, there are multiple signatures in proportion to the number of parallel paths similarly to the above statement.   □

5   Generic Construction on Network Graph

High Level Discussion of the Problem. Sequential aggregate signatures have no correctness on network graphs and overcoming the problem is difficult. Meanwhile, general aggregate signatures meets the correctness and hence we discuss how to fix the insecurity of them hereinafter. An essential problem of the insecurity is to amplify the value of signatures via network graphs. More specifically, for any aggregate signature scheme, signatures are written as $\prod_{j=1}^{i} \sigma_j^{\omega_j(\psi_i)}$. Whereas the attack depends on a kind of a group homomorphism, a group homomorphism is necessary for constructing aggregate signatures in order to combine individual signatures. In other words, the attack is due to an aggregation based on the group homomorphism, and is unavoidable as long as the homomorphism is used.

In order to overcome the problem, we propose two potential approaches, the weight function in small space and a homomorphism without the weight function. The former approach is instantiated by restricting the number of signers. A similar attack for breaking multisignatures has been shown in [33] and the restriction is their technique to prevent the attack. Since the attack scenario is to generate a graph whose weight is equal to zero under the modulo p, it can be avoided by checking values of the weight function. The upper bound of the weight function is given as $3^{\#\psi_i/3}$ as described in the previous section, and hence signers can resist the attack if the number of signers is less than the value. The other approach is to utilize a ring homomorphism. The ring homomorphism contains an addition and a multiplication, e.g., $+$ and $*$, and each operation is assigned to either one of a serial graph or a parallel graph. For instance, a signature equation can be obtained as $(\sigma_1 + \sigma_2) * \sigma_3$ and it does not contain the weight function. Therefore, this construction can resist the attack without restricting the number of signers in comparison with the former approach.

We note that our proposed schemes are generic conversions from aggregate signatures to secure ones: that is, a scheme to prevent the attack can be proposed if there exists an aggregate signature scheme with either one of the properties. We also emphasize that these proposed schemes are different form existing general aggregate signatures in a strict sense. In particular, verification and aggregation algorithms of our schemes contain graph structures and hence can include mechanisms to prevent the attack.

5.1   Construction with Restriction on the Number of Signers

Suppose that there exists an aggregate signature scheme with a group addition $+$, and we denote by (**Setup, KeyGen, Sign, Aggregate, Verify**) its algorithms. We then obtain the following construction. Here, we denote by (**SetupW, Key-GenW, SignW, AggregateW, VerifyW**) its algorithms. We also note that a message space is extended in the aggregation algorithm and the verification algorithm of our scheme.

SetupW Given a security parameter 1^k, generate a public parameter $para$ with **Setup**(1^k).

KeyGenW Given $para$, output a secret key sk_i and its corresponding public key pk_i by **KeyGen**$(para)$.

SignW Given $para$, sk_i and a message m_i, generate a signature σ_i by **Sign**$(para, sk_i, m_i)$ and then output σ_i. If any input is an error, then output an error symbol $\perp$.

AggregateW Given $para$ and i-tuples $(\{(pk_j, (m_j, \psi_j), \sigma'_j\}^i_{j=1})$ of public keys, messages consisting of ones to be signed and its graph and their signatures, iterate the following steps for $j \in [1, i]$: compose $\psi_{T(j)} := \cup_{k \in T(j)} \psi_k$; compute $\sigma_{T(j)} = \sum_{k \in T(j)} \sigma_k$ where set $\sigma_k = 0$ if $T(j)$ for any j is an empty, i.e., j is equal to one; compose $\psi_j := \psi_{T(j)} \cap \phi_j$; compute $\sigma_j = \sigma'_j + \sigma_{T(j)}$. If there are k signers $\{pk_k\}_{k \in T(i)}$ as the last signers of the whole signing group, then the whole structure is $\psi_i = \cup_{k \in T(i)} \psi_k$ and the whole signature is $\sigma = \sum_{k \in T(i)} \sigma_k$. Output a single short signature σ.

VerifyW Given $(para, \{pk_j, (m_j, \psi_j)\}^i_{j=1}, \sigma)$ as input, compose a whole graph ψ_i and checks if there exists an index j such that $\omega_j(\psi_i)$ for all j. If so, output $\perp$. Otherwise, run **Verify** with $(para, \{\omega_j(\psi_i)pk_j, m_j\}^i_{j=1}, \sigma)$. If the output is $\top$, output $\top$. If not, output $\perp$.

Theorem 3. Suppose that a general aggregate signature is $(t, q_c, q_s, q_h, n, \epsilon)$-secure. Then, the proposed scheme is $(t', q_c, q_s, q_h, n, \epsilon)$-secure, where $t' = t + \mathcal{O}(n)$ and $\mathcal{O}$ is the computational cost for generating a single signature.

Proof (Sketch). Due to the space limitation, we briefly describe the intuition of the security proof in this section. An intuition of the proof is that its signature distribution is almost the same as that of an ordinary aggregate signature scheme as long as the number of signers is restricted such that it is less than $3^{\#\psi_i/3}$. The goal of the proof is to forge the ordinary scheme via a forgery of the proposed scheme. In other words, constructing an adversary $\mathcal{B}$ against the ordinary scheme by the adversary $\mathcal{A}$ against the proposed scheme is the goal of the proof.

More specifically, $\mathcal{B}$ is given a challenge key from a challenger of the ordinary scheme, send it to $\mathcal{A}$ as a challenge key of the proposed scheme. When $\mathcal{A}$ gives a query to $\mathcal{B}$, $\mathcal{B}$ forwards the query to the challenger in order to respond to the query. The key generation algorithms and the signing algorithms are exactly identical for these games, and thus $\mathcal{B}$ can perfectly simulate the key certification query and the signing query. Likewise, for the random oracle queries, $\mathcal{B}$ can forward queries given from $\mathcal{A}$ to the challenger in order to simulate the random oracles. Since messages to be actually signed are identical to that of the ordinary scheme, the distribution of the random oracles are exactly identical.

After $\mathcal{A}$ outputs a forgery $(\{((m_i^*, \psi_i^*), pk_i^*)\}_{i=1}^n, \sigma^*)$, $\mathcal{B}$ excludes co-signers' signatures from the given forgery in order to obtain its target signer's signature $\omega_i(\psi_i^*)\sigma'$, where i means the index of the target signer. This exclusion can be executed since $\mathcal{B}$ knows all the co-signers' secret keys via the key certification queries. $\mathcal{B}$ computes σ' dividing by $\omega_i(\psi_n^*)$, and then reconstructs σ^* by utilizing the co-signers' secret keys. Finally, $\mathcal{B}$ outputs the reconstructed σ^* as a forgery.

The number of queries and the success probability of $\mathcal{B}$ is exactly the same as that of $\mathcal{A}$. However, for the given forgery from $\mathcal{A}$, $\mathcal{B}$ has to execute the signing algorithm for $(n-1)$ signers. Thus, $\mathcal{B}$'s execution time t' is that of $\mathcal{A}$ plus to the execution times on the signing algorithm at $(n-1)$ times. That is, $t' = t + \mathcal{O}(n)$ holds where $\mathcal{O}$ is the computational cost of the signing algorithm. $\square$

Although we omit instantiations of the scheme described above due to the page limitation, we can propose instantiations via the BGLS03 scheme [7], the GR06 scheme [13] or the HSW13 scheme [17].

5.2   Construction with Ring Homomorphism

Suppose that there exists a general aggregate signature scheme with a group addition $+$ and a group multiplication $*$, and we denote by (**Setup, KeyGen, Sign, Aggregate, Verify**) its algorithms. We then obtain the following construction. Here, we denote by (**SetupR, KeyGenR, SignR, AggregateR, VerifyR**) its algorithms. We also denote by $f(pk_i, h_i)$ a verification on Verification for all pk_i and all hash digest h_i on m_i, and note that a message space is extended in our scheme.

SetupR Given a security parameter 1^k, generate a public parameter *para* with **Setup**(1^k).
KeyGenR Given *para*, output a secret key sk_i and its corresponding public key pk_i by **KeyGen**(*para*).

SignR Given $para$, sk_i and a message m_i, generate a signature σ_i by **Sign**($para$, sk_i, m_i) and then output σ_i. If **Sign** outputs $\perp$, output $\perp$.

AggregateR Given $para$ and i-tuples ($\{(pk_j, (m_j, \psi_j), \sigma'_j\}^i_{j=1}$) of public keys, messages consisting of ones to be signed and its graph and their signatures, iterate the following steps for $j \in [1, i]$: compose $\psi_{\mathcal{T}(j)} := \cup_{k \in \mathcal{T}(j)} \psi_k$; compute $\sigma_{\mathcal{T}(j)} = \sum_{k \in \mathcal{T}(j)} \sigma_k$ where set $\sigma_k = 0$ if $\mathcal{T}(j)$ for any j is an empty, i.e., j is equal to one; compose $\psi_j := \psi_{\mathcal{T}(j)} \cap \phi_j$; compute $\sigma_j = \sigma'_j * \sigma_{\mathcal{T}(j)}$. If there are k signers $\{pk_k\}_{k \in \mathcal{T}(i)}$ as the last signers of the whole signing group, then the whole structure is $\psi_i = \cup_{k \in \mathcal{T}(i)} \psi_k$ and the whole signature is $\sigma = \sum_{k \in \mathcal{T}(i)} \sigma_k$. Output a single short signature σ.

VerifyR Given ($para$, $\{pk_j, (m_j, \psi_j)\}^i_{j=1}, \sigma$) as input, extract the verification function $f(pk_i, h_i)$ from **Verify** for all $j \in [1, j]$. Then, check if $f(\sigma, para) = f(pk_1, h_1) \odot_1 \cdots \odot_{i-1} f(pk_i, h_i)$ holds, where $\odot_j$ means $*$ if the relation among pk_j and pk_{j+1} is a parallel structure, otherwise, $*$. In addition, for the calculation of the right hand side of the congruence, $+$ operation is searched and any found $*$ operation is computed at first one by one until all $+$ operations are computed. Then all $*$ operations are computed one by one from the leftmost $*$ operation until the rightmost one. If the equation does not hold, output $\perp$. Otherwise, output $\top$.

Theorem 4. Suppose that a general aggregate signature is $(t, q_c, q_s, q_h, n, \epsilon)$-secure. Then, the proposed scheme is $(t', q_c, q_s, q_h, n, \epsilon)$-secure, where $t' = t + \mathcal{O}'(n)$ and $\mathcal{O}'$ is the costs of the signing algorithm and the inversion in a group homomorphism.

Proof (Sketch). Due to the page limitation, we give only intuition of the proof. Similarly to the proof in the previous section, its signature distribution is almost the same as that of an ordinary aggregate signature scheme. Thus, we can prove the security in a similar manner to the proof in the previous section.   □

Open Problem to Instantiation. We do not have instantiation of the construction described above. However, we consider that the existing homomorphic signatures such as the ALP13 scheme [2] and transitive signatures such as the CH12 scheme [9] are expected as instantiations. While these schemes are able to combine individually generated signatures, their combination algorithms do not take secret keys as input. It is close to the general aggregation. We leave as open problems to propose an instantiation of the proposed scheme.

6 Conclusion

Aggregate signatures are digital signatures for multiple signers where each signer can sign an individual document, and there are two kinds of aggregate signatures, general aggregate signatures and sequential aggregate signatures. Although one of key applications of aggregate signatures is routing protocols, we found potential vulnerabilities of both general aggregate signatures and sequential aggregate signatures. More specifically, we showed the unavailability of sequential aggregate signatures and the insecurity of general aggregate signatures on network graphs, which are mathematically idealized networks. We also proposed two approaches to overcoming the insecurity of general aggregate signatures. One of the approaches is to restrict the number of signers. Since an adversary constructs a malicious network graph in the attack we found, the success probability can be negligibly small by the restriction in the number of signers. The other approach is to construct an aggregate signature scheme based on a ring homomorphism. In such a scheme, an adversary is unable to construct a malicious network required for the attack. As a future work, we plan to construct an aggregate signature scheme based on a ring homomorphism. We also plan to implement our schemes.

Acknowledgement. Part of this research is supported by JSPS A3 Foresight Program. The first author is also supported by Support Center for Advanced Telecommunications Technology Research and JSPS KAKENHI Grant Numbers 26880012, 26330151. We would like to appreciate their supports. We would also like to appreciate Shin-Akarui-Angou-Benkyou-Kai for their valuable comments.

References

1. Ahn, J.H., Green, M., Hohenberger, S.: Synchronized aggregate signatures: new definitions, constructions and applications. In: Proceedings of CCS 2011, pp. 473–484. ACM (2010)
2. Attrapadung, N., Libert, B., Peters, T.: Efficient completely context-hiding quotable and linearly homomorphic signatures. In: Kurosawa, K., Hanaoka, G. (eds.) PKC 2013. LNCS, vol. 7778, pp. 386–404. Springer, Heidelberg (2013)
3. Bellare, M., Namprempre, C., Neven, G.: Unrestricted aggregate signatures. In: Arge, L., Cachin, C., Jurdziński, T., Tarlecki, A. (eds.) ICALP 2007. LNCS, vol. 4596, pp. 411–422. Springer, Heidelberg (2007)
4. Bodlaender, H.L., de Fluiter, B.: Parallel algorithms for series parallel graphs. In: Diaz, J., Serna, M. (eds.) ESA 1996. LNCS, vol. 1136, pp. 277–289. Springer, Heidelberg (1996)
5. Boldyreva, A., Gentry, C., O'Neill, A., Yum, D.H.: Ordered multisignatures and identity-based sequential aggregate signatures, with applications to secure routing (extended abstract). In: Proceedings of CCS, pp. 276–285. ACM (2007)
6. Boldyreva, A., Gentry, C., O'Neill, A., Yum, D.H.: Ordered multisignatures and identity-based sequential aggregate signatures, with applications to secure routing (extended abstract), (full paper) (2010)
7. Boneh, D., Gentry, C., Lynn, B., Shacham, H.: Aggregate and verifiably encrypted signatures from bilinear maps. In: Biham, E. (ed.) EUROCRYPT 2003. LNCS, vol. 2656, pp. 416–432. Springer, Heidelberg (2003)
8. Jain, A., Krenn, S., Pietrzak, K., Tentes, A.: Commitments and efficient zero-knowledge proofs from learning parity with noise. In: Wang, X., Sako, K. (eds.) ASIACRYPT 2012. LNCS, vol. 7658, pp. 663–680. Springer, Heidelberg (2012)
9. Camacho, P., Hevia, A.: Short Transitive Signatures for Directed Trees. In: Dunkelman, O. (ed.) CT-RSA 2012. LNCS, vol. 7178, pp. 35–50. Springer, Heidelberg (2012)

10. Camenisch, J.L., Lysyanskaya, A.: Signature schemes and anonymous credentials from bilinear maps. In: Franklin, M. (ed.) CRYPTO 2004. LNCS, vol. 3152, pp. 56–72. Springer, Heidelberg (2004)
11. Dingledine, R., Mathewson, N., Syverson, P.: Tor: the second-generation onion router. In: Proceedings of Usenix Security 2004 (2004)
12. Fischlin, M., Lehmann, A., Schröder, D.: History-free sequential aggregate signatures. In: Visconti, I., De Prisco, R. (eds.) SCN 2012. LNCS, vol. 7485, pp. 113–130. Springer, Heidelberg (2012)
13. Gentry, C., Ramzan, Z.: Identity-based aggregate signatures. In: Yung, M., Dodis, Y., Kiayias, A., Malkin, T. (eds.) PKC 2006. LNCS, vol. 3958, pp. 257–273. Springer, Heidelberg (2006)
14. Gerbush, M., Lewko, A., O'Neill, A., Waters, B.: Dual form signatures: an approach for proving security from static assumptions. In: Wang, X., Sako, K. (eds.) ASIACRYPT 2012. LNCS, vol. 7658, pp. 25–42. Springer, Heidelberg (2012)
15. Giesen, F., Kohlar, F., Stebila, D.: On the security of tls renegotiation. In: Proceedings of CCS 2013, pp. 387–398. ACM (2013)
16. Hohenberger, S., Koppula, V., Waters, B.: Universal signature aggregators. In: Oswald, E., Fischlin, M. (eds.) EUROCRYPT 2015. LNCS, vol. 9057, pp. 3–34. Springer, Heidelberg (2015)
17. Hohenberger, S., Sahai, A., Waters, B.: Full domain hash from (leveled) multilinear maps and identity-based aggregate signatures. In: Canetti, R., Garay, J.A. (eds.) CRYPTO 2013, Part I. LNCS, vol. 8042, pp. 494–512. Springer, Heidelberg (2013)
18. Hohenberger, S., Sahai, A., Waters, B.: Replacing a random oracle: full domain hash from indistinguishability obfuscation. In: Nguyen, P.Q., Oswald, E. (eds.) EUROCRYPT 2014. LNCS, vol. 8441, pp. 201–220. Springer, Heidelberg (2014)
19. Hu, Y.C., Perrig, A., Johnson, D.B.: Ariadne: a secure on demand routing protocol for ad hoc network. Wireless Netw. 11, 21–38 (2005)
20. Hwang, J.Y., Lee, D.H., Yung, M.: Universal forgery of the identity-based sequential aggregate signature scheme. In: Proceedings of ASIACCS, pp. 157–160. ACM (2009)
21. Kent, S., Lynn, C., Seo, K.: Secure border gateway protocol. IEEE J. Sel. Areas Commun. 18(4), 582–592 (2000)
22. Krawczyk, H., Paterson, K.G., Wee, H.: On the security of the TLS protocol: a systematic analysis. In: Canetti, R., Garay, J.A. (eds.) CRYPTO 2013, Part I. LNCS, vol. 8042, pp. 429–448. Springer, Heidelberg (2013)
23. Lee, K., Lee, D.H., Yung, M.: Aggregating CL-signatures revisited: extended functionality and better efficiency. In: Sadeghi, A.-R. (ed.) FC 2013. LNCS, vol. 7859, pp. 171–188. Springer, Heidelberg (2013). http://fc13.ifca.ai/proc/5-2.pdf
24. Lee, K., Lee, D.H., Yung, M.: Sequential aggregate signatures with short public keys: design, analysis and implementation studies. In: Kurosawa, K., Hanaoka, G. (eds.) PKC 2013. LNCS, vol. 7778, pp. 423–442. Springer, Heidelberg (2013)
25. Lepinski, M., Turner, S.: An overview of bgpsec, October 2011. Internet Draft. http://tools.ietf.org/html/draft-ietf-sidr-bgpsec-overview-01
26. Lu, S., Ostrovsky, R., Sahai, A., Shacham, H., Waters, B.: Sequential aggregate signatures and multisignatures without random oracles. In: Vaudenay, S. (ed.) EUROCRYPT 2006. LNCS, vol. 4004, pp. 465–485. Springer, Heidelberg (2006)
27. Lysyanskaya, A., Micali, S., Reyzin, L., Shacham, H.: Sequential aggregate signatures from trapdoor permutations. In: Cachin, C., Camenisch, J.L. (eds.) EUROCRYPT 2004. LNCS, vol. 3027, pp. 74–90. Springer, Heidelberg (2004)
28. Neven, G.: Efficient sequential aggregate signed data. IEEE Trans. Inf. Theor. 57(3), 1803–1815 (2011)

29. Rekhter, Y., Li, T.: A border gateway protocol 4 (bgp-4). RFC 1771, March 1995. http://www.ietf.org/rfc/rfc1771.txt
30. Rückert, M., Schröder, D.: Aggregate and verifiably encrypted signatures from multilinear maps without random oracles. In: Park, J.H., Chen, H.-H., Atiquzzaman, M., Lee, C., Kim, T., Yeo, S.-S. (eds.) ISA 2009. LNCS, vol. 5576, pp. 750–759. Springer, Heidelberg (2009)
31. Schröder, D.: How to aggregate the CL signature scheme. In: Atluri, V., Diaz, C. (eds.) ESORICS 2011. LNCS, vol. 6879, pp. 298–314. Springer, Heidelberg (2011)
32. Sun, Y., Edmundson, A., Vanbever, L., Li, O., Rexford, J., Chiang, M., Mittal, P.: Raptor: routing attacks on privacy in tor. In: Proceedings of Usenix Security 2015, pp. 271–286 (2015)
33. Tada, M.: A secure multisignature scheme with signing order verifiability. IEICE Trans. Fundam. Electron. Commun. Comput. Sci. **86**(1), 73–88 (2003)
34. Valera, F., Beijnum, I.V., Garcia-Martinez, A., Bagnulo, M.: Multi-path BGP: Motivations and Solutions, Chapter 1, pp. 238–256. Cambridge University Press, Cambridge (2011)
35. Waters, B.: Efficient identity-based encryption without random oracles. In: Cramer, R. (ed.) EUROCRYPT 2005. LNCS, vol. 3494, pp. 114–127. Springer, Heidelberg (2005)
36. Waters, B.: Dual system encryption: realizing fully secure IBE and HIBE under simple assumptions. In: Halevi, S. (ed.) CRYPTO 2009. LNCS, vol. 5677, pp. 619–636. Springer, Heidelberg (2009)
37. Zhao, M., Smith. S., Nicol, D.: Aggregated path authentication for efficient bgp security. In: Proceedings of CCS, pp. 128–138. ACM (2005)

Universally Composable Oblivious Database in the Presence of Malicious Adversaries

Huafei Zhu[⊠]

School of Computer and Computing Science, Zhejaing University City College,
Hangzhou 310015, China
zhuhf@zucc.edu.cn

Abstract. The core technique for constructing oblivious database is to get efficient implementations of oblivious transfer. This paper studies universally composable 1-out-of-n oblivious transfer (OT_1^n) in the presence of malicious adversaries under the standard cryptographic assumptions. Our oblivious transfer protocol is constructed from the Damgård and Jurik's double trapdoor encryption scheme and the Damgård and Nielsen's mixed commitment scheme, where the master key of the underlying double trapdoor cryptosystem is used to extract implicit input of a corrupted sender while the corresponding local keys are used to extract implicit input of a corrupted receiver. We claim that the proposed oblivious transfer framework realizes the universally composable security in the common reference model under the joint assumptions that the decisional Diffie-Hellman problem and the decisional composite residuosity problem are hard as well as all knowledge proof protocols applied are zero-knowledge.

Keywords: Double trapdoor cryptosystem · Mixed commitment scheme · Oblivious transfer · Universal composability

1 Introduction

Nowadays, user data is becoming a new economic asset that will touch all aspects of our modern society. User data collected from different resources is stored in a user-controlled data box which in essence, is a cloud storage associated with a specified access control strategy. The collected data stored in the personal box can be sold to a data broker (say, Bob). In return, the user (say, Alice) gets the repayment from Bob. The broker Bob manages and maintains the collected data and provides the data-as-a-service leveraging the established database systems in the client-server model, where the broker Bob serves as a database management server. One of such services is oblivious information retrieval where the client should obtain correct answer to the query and nothing else is leaked while the server should not learn anything besides the type of the query. Commonly, a database system supports the oblivious information retrieval procedure is referred to as an oblivious database [14, 20, 21].

© Springer International Publishing Switzerland 2016
M. Yung et al. (Eds.): INTRUST 2015, LNCS 9565, pp. 49–61, 2016.
DOI: 10.1007/978-3-319-31550-8_4

The core technique for constructing oblivious database is to get efficient implementations of oblivious transfer (OT) which was first introduced by Rabin [29], and generalized by Even et al. [16] and Brassard et al. [2]. Typically, an OT database comprises two parties: a sender (or a server) and a receiver (or a customer), where a sender has input messages $m_1, \ldots, m_n$, interactively communicating with a receiver with a specified index $\sigma \in [1, ..., n]$ in such a way that at the end of the protocol execution, the receiver obtains m_σ while the sender learns nothing. OT stands at the center of the fundamental results on secure two-party and multi-party computation showing that any efficient functionality can be securely computed [7,19,31]. Due to its general importance, the task of constructing efficient oblivious transfer protocols has attracted much interest (see [1,3,9,10,12,20,22–26,32,33] for more details).

1.1   Motivation Problem

At Crypto'08, Peikert et al. [28] proposed a general framework of universally composable 1-out-of-2 oblivious transfer protocols in the presence of static adversaries under the standard cryptographic assumptions. Their protocols are based on a new notion called dual-mode cryptosystem. Such a system starts with a setup phase that produces a common reference string, which is made available to all parties. The cryptosystem is set up in one of two modes: extraction mode or decryption mode. A crucial property of the dual-mode cryptosystem is that no adversary can distinguish the common reference string between two modes. To prove the security against a malicious sender, a simulator must run a trapdoor extractable algorithm that given a trapdoor t, outputs (pk, sk_0, sk_1), where pk is a public encryption key and sk_0 and sk_1 are corresponding secret keys for index 0 and 1 respectively. To prove the security against a malicious receiver, a simulator must run a find-lossy algorithm that given a trapdoor t and pk, outputs an index corresponding to the message-lossy index of pk.

Recall that an efficient construction of 1-out-of-4 oblivious transfer protocols is a key start-up for secure multi-arty computations in the framework of Goldreich, Micali and Wigderson. As a result, if Alice is given an efficient implementation of the 1-out-of-4 OT protocol (e.g., a construction based on the Peikert, Vaikuntanathan and Waters' general framework under the standard cryptographic assumption), then she is able to securely compute any function f in terms of Boolean circuits [10]. We however, do not know how the Peikert et al's general framework for 1-out-of-2 oblivious transfer protocols can be used to construct 1-out-of-4 (and the general case 1-out-of-n) oblivious transfer protocols under the standard cryptographic assumptions which leaves the following interesting research problem:

How to construct universally composable 1-out-of-n oblivious transfer protocols in the presence of static adversaries under the standard cryptographic assumptions?

1.2   This Work

This paper provides a construction of universally composable 1-out-of-n oblivious transfer (OT_1^n) in the presence of static adversaries under the standard cryptographic assumptions based on the Damgård and Jurik's double-trapdoor encryptions and the Damgård and Nielsen's mixed commitment schemes. Using the oblivious transfer extension technique [18], we are able to to establish a large oblivious database system where millions of oblivious transfers are invoked for an information retrieval processing by applying the base OT_1^n constructed in this paper.

Overview of the Implementation. Our construction comprises three phases: a common reference string generation phase, an initialization phase and an interactive communication phase.

- In the common reference string generation phase, two reference strings are generated: a common reference string generated from the Damgård and Jurik's double-trapdoor encryption [15] and a common reference string generated from the Damgård and Nielsen's mixed commitment scheme [13]. The common reference string of the mixed commitment allows a simulator to interpret a dummy commitment as a commitment of any message. The common reference string of the double-trapdoor cryptosystem allows the simulator to extract implicit input of a corrupted sender with the help of the master secret key. The common reference string also allows the simulator to extract implicit input of a corrupted receiver with the help of the local secret keys.
- In the initialization phase, the sender will set up a committed database by means of the mixed commitment scheme. A crucial property of the committed database is that the committed database is extractable when the sender gets corrupted and the committed database is equivocable when the receiver gets corrupted.
- In the interactive communication phase, the receiver retrieves messages in such a way that at the end of the protocol execution, the receiver obtains a message m_σ while the sender learns nothing.

Overview of the Proof. Let (mpk, pk_1, ..., pk_n) be a common reference derived from the Damgård and Jurik's double-trapdoor encryption and (msk, sk_1, ..., sk_n) be the corresponding secret keys, where (mpk, msk) is a pair of master public and master secret keys and (pk_i, sk_i) is a pair of local keys. Given ($mpk, pk_1, ..., pk_n$), an ideal-world adversary works as follows: to prove receiver's security against a malicious sender, the master secret key msk is used to extract implicit input of the corrupted sender. Consequently, a simulator for the corrupted sender is well defined; to prove sender's security against a malicious receiver, the local keys $sk_1, ..., sk_n$ are used to extract implicit input of the corrupted receiver and then the simulator makes use of the equivocal keys to interpret a fake commitment to a commitment of any message. Such an interpretation is necessary when the simulator does not have any knowledge of the honest sender. In our construction, zero-knowledge proof technique such as the Cramer-Damgård-Schoenmakers' OR-protocol [6] for proving the OR-relationship of Diffie-Hellman quadruples are used. We stress here that we do NOT apply the rewinding technique for extracting input messages of corrupted parties. The simulator extracts the inputs messages of corrupted parties by means of the master key of the double trapdoor encryption scheme. This is our key idea to construct 1-out-of-n composable oblivious transfers.

Main Result. We claim that the proposed oblivious transfer protocol OT_1^n realizes the universally composable security in the common reference model under the joint assumptions that the decisional Diffie-Hellman problem and the decisional composite residuosity problem are hard as well as all knowledge proof protocols applied in our implementation are zero-knowledge in the presence of static adversaries.

Road Map. The rest of this paper is organized as follows: The security notion of 1-out-of-n oblivious transfer protocols in the UC-framework and the building blocks are presented in Sect. 2. A detailed description of 1-out-of-n oblivious transfer protocol is presented in Sect. 3 and a proof of its security is presented in Sect. 4. We conclude this work in Sect. 5.

2 Preliminaries

We assume that the reader is familiar with the standard notion of UC-Security (we refer to the reader [4] for a detailed description of the executions, and definitions of $\mathrm{IDEAL}_{\mathcal{F},\mathcal{S},\mathcal{Z}}$ and $\mathrm{REAL}_{\pi,\mathcal{A},\mathcal{Z}}$).

Common Reference String Model. Canetti and Fischlin have shown that OT cannot be UC-realized without a trusted setup assumption [8]. We thus assume the existence of an honestly-generated Common Reference String (crs) and work in the so called $\mathcal{F}_{crs}^{\mathcal{D}}$-hybrid model. The functionality of common reference string model [5] assumes that all participants have access to a common string that is drawn from some specified distribution $\mathcal{D}$. The details follow:

Functionality $\mathcal{F}_{crs}^{\mathcal{D}}$

$\mathcal{F}_{crs}^{\mathcal{D}}$ proceeds as follows, when parameterized by a distribution $\mathcal{D}$.

When receiving a message (sid, P_i, P_j) from P_i, let $\mathsf{crs} \leftarrow \mathcal{D}(1^n)$ and send $(\mathsf{sid}, \mathsf{crs})$ to P_i, and send $(\mathsf{sid}, P_i, P_j, \mathsf{crs})$ to the adversary, where sid is a session identity.

Next when receiving (sid, P_i, P_j) from P_j (and only from P_j), send $(\mathsf{sid}, \mathsf{crs})$ to P_j and to the adversary, and halt.

Functionality for 1-out-of-n Oblivious Transfer OT_1^n. The description of the functionality $\mathcal{F}_{\mathrm{OT}_1^n}$ [7] follows:

Functionality $\mathcal{F}_{\text{OT}_1^n}$

$\mathcal{F}_{\text{OT}_1^n}$ proceeds as follows, parameterized with κ and n, and running with an oblivious transfer sender S, a receiver R and an ideal world adversary $\mathcal{S}$.

- Upon receiving a message (sid, sender, $m_1, \ldots, m_n$) from S, where each $m_i \in \{0,1\}^\kappa$, an imaginary trusted third party (TTP) stores $(m_1, \ldots, m_n)$;
- Upon receiving a message (sid, receiver, σ) from R, TTP checks if a (sid, sender, $\ldots$) message was previously received. If no such message was received, TTP sends nothing to R. Otherwise, TTP sends (sid, request) to S and receives the tuple (sid, $b \in \{0,1\}$) in response.
- TTP then passes (sid, b) to the adversary, and: if $b = 0$, TTP sends (sid, $\perp$) to R; if $b = 1$, TTP sends (sid, m_σ) to R.

Definition 1. *Let $\mathcal{F}$ be a functionality for OT_1^n. A protocol π is said to universally composably realize $\mathcal{F}$ if for any adversary $\mathcal{A}$, there exists a simulator $\mathcal{S}$ such that for all environments $\mathcal{Z}$, the ensemble $\text{IDEAL}_{\mathcal{F},\mathcal{S},\mathcal{Z}}$ is computationally indistinguishable with the ensemble $\text{REAL}_{\pi,\mathcal{A},\mathcal{Z}}$.*

Mixed Commitments. A mixed commitment (MC) scheme [13] is a commitment scheme with a global public key gpk which determines the message space $\mathcal{M}_c$ and key space $\mathcal{K}_c$ of the commitments. A mixed commitment scheme has three key generation algorithms (GenMC, EKey, XKey). On input a security parameter 1^k, the global key generation algorithm GenMC outputs a pair of keys (gpk, spk). On input gpk, the extractable key generation algorithm XKey outputs a random extractable key K; on input gpk, the equivocal key generation algorithm EKey outputs a pair of keys (K, τ_K). A mixed commitment scheme has to satisfy the next three properties:

- Key indistinguishability: Random equivocal keys and random extractable keys are both computationally indistinguishable from random keys (a random key K is an element chosen uniformly at random from the key space $\mathcal{K}_c$, i.e., $K \leftarrow_R \mathcal{K}_c$) as long as the global secret key gsk is not known;
- Equivocability: Given equivocal key K and corresponding trapdoor τ_K, one can generate a fake commitment distributed as real commitments, which can later be open arbitrarily, i.e., given a message m, one can compute a random-looking r for which $c = \text{Com}_K(m, r)$;
- Extraction: Given a commitment $c = \text{Com}_K(m, r)$, where K is an extractable key, one can efficiently compute m given gsk, where m is uniquely determined by the perfect binding.

The indistinguishability of random equivocal keys, random extractable keys and random keys implies that as long as the global secret key gsk is unknown, then mixed commitment is computationally hiding for all keys and as long as neither the global secret key gsk nor the equivocal trapdoor τ_K is known, the mixed commitment is computationally binding for all keys.

Damgård-Nielsen Mixed Commitment Scheme [13]. Let $\tilde{N} = \tilde{p}\tilde{q}$ be an RSA modulus. Let $gpk = \tilde{N}$ and $gsk = (\tilde{p}, \tilde{q})$. Let $s \geq 0$ be some fixed constant.

The key space is $\mathbb{Z}_{\tilde{N}^{s+1}}$, the message space is $\mathbb{Z}^*_{\tilde{N}^s}$, and the randomness space is $\mathbb{Z}^*_{\tilde{N}}$. Define $\psi(m,r) := (1 + \tilde{N})^m r^{\tilde{N}^s} \mod N^{s+1}$. Equivocal keys are elements of form $\psi(0,r)$ and the trapdoor is $r \in \mathbb{Z}^*_{\tilde{N}}$. Extractable keys are of form $\psi(k,r_k)$, where $k \in \mathbb{Z}_{\tilde{N}^s}$ and $r_k \in Z^*_N$. Given a key $K \in \mathbb{Z}^*_{N^{s+1}}$ and a message $m \in \mathbb{Z}_{\tilde{N}^s}$, let r be a uniformly random element from $\mathbb{Z}^*_{\tilde{N}}$ and commit as $c = \mathsf{Com}_K(m,r) = K^m r^{N^s} \mod \tilde{N}^{s+1}$. The commitment scheme defined by $\mathsf{Com}_K(m,r) = K^m r^{N^s} \mod \tilde{N}^{s+1}$ is a mixed commitment scheme, assuming that the decisional composite residuosity problem is hard [13].

Damgård-Jurik's Double-trapdoor Cryptosystem. Damgård and Jurik [15] presented a length-flexible public-key cryptosystem derived from the Paillier's encryption. Their cryptosystem is length-flexible in the sense that public-key can be set up so that messages of arbitrary length can be handled efficiently with the same fixed keys. The Damgård-Jurik cryptosystem [15] consists of the following three algorithms.

- Key generation algorithm $\mathsf{DTGen} = (\mathsf{DTGen}_{\mathsf{master}}, \mathsf{DTGen}_{\mathsf{partial}})$: On input a security parameter λ, the master key generation algorithm $\mathsf{DTGen}_{\mathsf{master}}$ outputs p, q and N, where N be a product of two large safe primes p and q p $=2p' + 1$ and $q = 2q' + 1$. Let $N' = p'q'$ and g be a random generator of QR_N. Let $h = g^x$, $x \in_R \mathbb{Z}_{N'}$. The master public key mpk is N and the master secret key msk is (p,q). The local public key pk is (g,h) and the local secret key sk is x.
- Encryption algorithm DTEnc: Given a message $m \in \mathbb{Z}^+$, a sender S chooses an integer $s > 0$ such that $m \in \mathbb{Z}_{N^s}$, and a random $r \in \mathbb{Z}_N$ and computes $u = g^r \mod n$, $v = (1 + N)^m (h^r \mod N)^{N^s} \mod N^{s+1}$. Let (u, v) be a ciphertext of the message m, i.e., $c = (u, v) = \mathsf{DTEnc}(m,r)$.
- Decryption algorithm DTDec: Given a ciphertext $c=(u, v)$, DTDec deduces s from the length of c (or retries s from the attached encryption) and then decrypts using the following two procedures: (1) the master key decryption procedure: The decryption algorithm DTDec first computes $v^{2N'} \mod N^{s+1}$ to obtain $(1 + N)^{m2N'} \mod N^s$, then extracts the message $m \in \mathbb{Z}_N$ from $(1 + N)^{m2N'} \mod N^s$; and (2) the partial key decryption procedure: The decryption algorithm DTDec first computes $(u^x \mod N)^{N^s} \mod N^{s+1}$ to obtain $(1 + N)^m \mod N^{s+1}$ and then extracts the message $m \in \mathbb{Z}_{N^s}$ from $(1 + N)^m \mod N^{s+1}$.

Under the hardness of the decisional composite residuosity and the decisional Diffie-Hellman problem in QR_N, the Damgård and Jurik double trapdoor cryptosystem is semantically secure [15].

Zero-knowledge Proof Protocols: Throughout the paper, we assume that the reader is familiar with the Cramer-Damgård-Schoenmakers' OR-protocol [6] for proving the equality of a commitment C_i and an encryption E_i ($i = 1, \ldots, n$).

3   A Construction of UC-secure OT_1^n

Our universally composable 1-out-n oblivious transfer protocol π described below consists of the following three phases (for simplicity, we assume that $s = 2$ in the Damgård and Jurik protocol and the Damgård-Nielsen mixed commitment protocol, one can extend the scheme to the general case where $s > 2$): a common reference generation phase, an initialization phase an OT-query phase.

The Common Reference String Generation Algorithm. A common reference string in our implementation consists of two reference strings: a reference string crs_{MC} for Damgård-Nielsen's mixed commitment scheme and a reference string crs_{DE} for Damgård-Jurik's Cryptosystem. The common reference string for the mixed commitment scheme enables a simulator to interpret a dummy ciphertext generated in the initialization protocol to a message that the corrupted receiver obtains during the course of the OT query protocol. A common reference string for the double trapdoor cryptosystem enables a simulator to extract an implicit input of corrupted sender. The details of the common reference string generation algorithm are depicted below:

- Given a security parameter λ, the common reference string generation algorithm $OTcrsGen(1^\lambda)$ generates composite modulus of the form $N = pq$ that is a product of two safe primes p and q. $OTcrsGen(1^\lambda)$ outputs a cyclic group $QR_N \subseteq \mathbb{Z}_N^*$ of order N' and n random generators $g_1, \ldots, g_n$ of QR_N. $OTcrsGen(1^\lambda)$ randomly chooses $x_1, \ldots, x_n \in \mathbb{Z}_{N'}$ and sets $h_i \leftarrow g_i^{x_i} \mod N$ ($i = 1, \ldots, n$). Let $mpk = N$ and $pk_i = (g_i, h_i)$. Let $crs_{DE} = < mpk, pk_1, \ldots, pk_n >$.
- $OTcrsGen(1^\kappa)$ then invokes the key generation algorithms (GenMC, XKey) of Damgård-Nielsen's mixed commitment scheme to generate the global public key $\tilde{N}$ and n Xkeys $K_1, \ldots, K_n \leftarrow_R \mathbb{Z}_{\tilde{N}^2}^*$, where $\tilde{N} = \tilde{p}\tilde{q}$, $\tilde{p} = 2\tilde{p}' + 1$, $\tilde{q} = 2\tilde{q}' + 1$, and $\tilde{p}$, $\tilde{p}'$, $\tilde{q}$ and $\tilde{q}'$ are large prime numbers. Let $crs_{MC} = \langle \tilde{N}, K_1, \ldots, K_n \rangle$.
- Let $crs = (crs_{DE}, crs_{MC})$. The common reference string generation algorithm $OTcrsGen(1^\lambda)$ broadcasts crs to all participants.

INITIALIZATION PROTOCOL. The goal of initialization protocol is to construct a committed database based on the Damgård-Nielsen's mixed commitment scheme. On input $(m_1, \ldots, m_n)$ ($m_i \in \mathcal{M}_c$) and $crs = (crs_{DE}, crs_{MC})$, the sender S performs the following computations ($1 \leq i \leq n$):

1. Upon receiving the instruction $(sid, sender, m_1, \ldots, m_n)$, S generates n commitments $C_j \leftarrow Com_{K_j}(m_j, r_j)$, where $r_j \leftarrow_R \mathcal{R}_c$. Let $C = (C_1, \ldots, C_n)$.
2. S broadcasts C to all participants.

THE OT_1^n PROCEDURE. The OT_1^n procedure consists of the following steps:

1. Given $crs_{DE} = (des(G), (pk_1, \ldots, pk_n), mpk)$, $crs_{MC} = (gpk, K_1, \ldots, K_n)$ and $\sigma \in [1, n]$, the receiver R parses pk_σ as (g_σ, h_σ) and takes pk_σ as input to generate a temporary public-key $tpk = (g_\sigma^{z_\sigma} \mod N, h_\sigma^{z_\sigma} \mod N)$, where $z_\sigma \leftarrow_R \mathbb{Z}_\beta$, $len(\beta) = 2\ell(\lambda)$ and $\ell(\lambda)$ is an upper bound of the bit length of

group order N'. The corresponding temporary secret key tsk is z_σ. R keeps tsk secret.

Let $(g, h) = (g_\sigma^{z_\sigma}, h_\sigma^{z_\sigma})$, and $U_i = (g_i, h_i, g, h)$. R then sends (sid, tpk, S, R) to S.

2. On input the common reference string crs and the temporary public-key tpk, the sender S performs the following computations:

   – Parsing tpk as (g, h), S takes OTcrsGen and (g, h) as input and checks the validity of the temporary public-key. S then checks that (1) $g \in Z_{N^2}^*$ and $h \in Z_{N^2}^*$; (2) $g^{2N} \neq 1$ and $h^{2N} \neq 1$. If any of two conditions are violated, then outputs $\perp$; otherwise, S outputs $\perp$.

   – If all the checks are valid, S takes $(pk_1, \ldots, pk_n)$ and tpk as input and generates a randomized partial public key rpk_i for the double trapdoor encryption algorithm DTEnc. That is, on input (g_i, h_i) and (g, h), S chooses $s_i, t_i \leftarrow_R \mathbb{Z}_\beta$ uniformly at random and outputs the randomized temporary public key $rpk_i \leftarrow (g_i^{s_i} h_i^{t_i}, g^{s_i} h^{t_i})$, for $i = 1, \ldots, n$.

   – Let $(u_i, v_i) = \mathsf{DTEnc}(rpk_i, m_i)$ and $E_i = (u_i, v_i)$, where $u_i = g_i^{s_i} h_i^{t_i} \bmod N$ and $v_i = (g^{s_i} h^{t_i})^N (1+N)^m \bmod N^2$. S then proves in zero-knowledge the following statements

   (1) Each E_i is a correctly generated ciphertext ($i = 1, \ldots, n$): Let $\tilde{u}_i = u_i^N \bmod N$ and $\tilde{v}_i = v_i \bmod N$. S proves that the exponential components (s_i, t_i) of $\tilde{v}_i$ on the bases $(g^N \bmod N, h^N \bmod N)$ equals the exponential components (s_i, t_i) of $\tilde{u}_i$ on the bases $(g_i^N \bmod N, h_i^N \bmod N)$. Let ZK-PKoE be a zero-knowledge proof of the above proof.

   (2) E_i and C_i hide the same message m_i: $< ((E_1, C_1)$ hiding the same message $m_1) \vee ((E_2, C_2)$ hiding the same message $m_2) \vee \ldots \vee ((E_n, C_n)$ hiding the same message $m_n) >$. Let ZK-PKoEQ be a zero-knowledge proof of the above OR-relationship from the Cramer, Damgård and Schoenmakers' protocol [6].

   (3) Finally, S sends $(E_1, \ldots, E_n)$, together with ZK-PKoE and ZK-PKoEQ to R.

3. Upon receiving $(sid, E_1, \ldots, E_n)$, ZK-PKoOR and ZK-PKoEQ, R checks the validity of the received PKoEQ. If the proofs ZK-PKoOR and ZK-PKoEQ are valid, R recovers $m_\sigma \leftarrow \mathsf{DTDec}(tsk, E_i)$; otherwise, R outputs $\perp$.

4   Proof of Security

Theorem 1. *The proposed oblivious transfer protocol OT_1^n is universally composable in the $\mathcal{F}_{crs}^{\mathcal{D}}$-hybrid model under the joint assumptions that the decisional Diffie-Hellman problem and the decisional composite residuosity problem are hard as well as all knowledge proof protocols applied are zero-knowledge in the presence of static adversaries.*

Proof. Let π be the described OT_1^n protocol. Let $\mathcal{A}$ be a static adversary that interacts with the parties S and R running the protocol π. We will construct an ideal world adversary $\mathcal{S}$ interacting with the ideal functionality $\mathcal{F}_{\mathrm{OT}}$ such that no environment $\mathcal{Z}$ can distinguish an interaction with $\mathcal{A}$ in the protocol π from an interaction with the simulator $\mathcal{S}$ in the ideal world. The construction of $\mathcal{S}$ depends on which of the two parties gets corrupted, and thus we will separately handle all different possibilities.

Simulating the case where only the sender S is corrupted. When S is corrupted and R is honest, the adversary $\mathcal{A}$ extracts S's input from the given commitments and then forwards the extracted messages to the OT_1^n functionality. The goal of an simulator $\mathcal{S}$ is to generate the remaining messages (namely, messages from R) so that the entire transcript is indistinguishable from the real interaction between S and R. The details of $\mathcal{S}$ are described below:

1. when the adversary $\mathcal{A}$ queries to $\mathsf{OTcrsGen}(1^\kappa)$ for a common reference string crs, the simulator invokes the key generation algorithm DTGen of the Damgård-Jurik's double trapdoor encryption scheme to generate composite modulus of the form $N = pq$ that is a product of two safe primes p and q (i.e., $p = 2p' + 1$, $q = 2q' + 1$), a cyclic group $G \subseteq Z_N^*$ of order N' ($N' = p'q'$), and n random generators $g_1, \dots, g_n$ of G and n random elements $(h_1, \dots, h_n)$ such that $h_i = g_i^{x_i} \bmod N$, $x_i \in_U Z_{N'}$, $i = 1, \dots, n$. The simulator keeps the auxiliary strings (p, q) and $(x_1, \dots, x_n)$ secret. Let $\mathsf{crs}_{\mathrm{DE}} = \ <\mathrm{des}(G), (g_1, h_1),$ $\dots, (g_n, h_n), N >$.
   $\mathsf{OTcrsGen}$ then invokes the key generation algorithms $(\mathsf{GenMC}, \mathsf{XKey})$ of Damgård-Nielsen's mixed commitment scheme to generate a global public key $\widetilde{N}$ and n extractable keys $K_1, \dots, K_n \in_U Z_{N^2}^*$, where $\widetilde{N} = \widetilde{p}\widetilde{q}$, $\widetilde{p} = 2\widetilde{p'} + 1$, $\widetilde{q} = 2\widetilde{q'} + 1$, and $\widetilde{p}, \widetilde{p'}, \widetilde{q}$ and $\widetilde{q'}$ are large prime numbers. Let $K_i = (1 + \widetilde{N})^{k_i} r_{k_i}^{\widetilde{N}} \bmod \widetilde{N}$ ($i = 1, \dots, n$). The simulator keeps $(\widetilde{p}, \widetilde{q})$ and $< (k_1, r_{k_1}), \dots, (k_n, r_{k_n}) >$ secret. Let $\mathsf{crs}_{\mathrm{MC}} = \ <\widetilde{N}, K_1, \dots, K_n >$.
   Let $\mathsf{crs} = (\mathsf{crs}_{\mathrm{DE}}, \mathsf{crs}_{\mathrm{MC}})$. $\mathcal{F}_{\mathrm{crs}}^D$ returns $(\mathsf{sid}, \mathsf{crs})$ to the adversary $\mathcal{A}$;
2. When the simulator $\mathcal{S}$ receives (sid, C) from the real world adversary $\mathcal{A}$ who fully controls the corrupted sender S, the simulator $\mathcal{S}$ extracts the messages $(m_1, \dots, m_n)$ from the given ciphertexts $(C_1, \dots, C_n)$ using the trapdoor strings $(\widetilde{p}, \widetilde{q})$ and $< (k_1, r_{k_1}), \dots, (k_n, r_{k_n}) >$ and then forwards the extracted messages $(m_1, \dots, m_n)$ to the ideal functionality $\mathcal{F}_{\mathrm{OT}_1^n}$. We stress that the simulator $\mathcal{S}$ must send the implicit messages $(m_1, \dots, m_n)$ to the functionality $\mathcal{F}_{\mathrm{OT}_1^n}$ at the initial stage.
3. $\mathcal{S}$ randomly selects (g, h) with order N' and sets $tpk = (g, h)$. We stress that the choice of (g, h) is a trivial task since $\mathcal{S}$ holds the master key (p, q). The simulator then sends $tpk = (g, h)$ to the adversary $\mathcal{A}$.
4. Upon receiving $(E_1, \dots, E_n)$, the transcripts ZK-PKoE and ZK-PKoEQ from the adversary $\mathcal{A}$, $\mathcal{S}$ checks the given proofs. If all checks are valid, $\mathcal{S}$ decrypts E_σ to reveal m_σ, otherwise, outputs $\bot$.

Let IDEAL$_{\mathcal{F},\mathcal{S},\mathcal{Z}}$ be the view of ideal world adversary $\mathcal{S}$ and REAL$_{\pi,\mathcal{A},\mathcal{Z}}$ be the view of real world adversary $\mathcal{A}$ according to the description of protocol π. The only difference between REAL$_{\pi,\mathcal{A},\mathcal{Z}}$ and IDEAL$_{\mathcal{F},\mathcal{S},\mathcal{Z}}$ is the generation of public key (g,h) in π and the public key (g,h) in the simulation procedure. Since $(g_\sigma, h_\sigma, g, h)$ is a random quadruple in IDEAL$_{\mathcal{F},\mathcal{S},\mathcal{Z}}$ while $(g_\sigma, h_\sigma, g, h)$ is a Diffie-Hellman quadruple in REAL$_{\pi,\mathcal{A},\mathcal{Z}}$. By the decisional Diffie-Hellman assumption, we know that REAL$_{\pi,\mathcal{A},\mathcal{Z}} \approx$ IDEAL$_{\mathcal{F},\mathcal{S},\mathcal{Z}}$.

Simulating the case where only the receiver R is corrupted. When R gets corrupted, the adversary generates all the messages from R. The goal of the simulator $\mathcal{S}$, then, is to generate the remaining messages (namely, all messages from S) so that the entire transcript is indistinguishable from the real interaction between S and R. The details of an ideal-world adversary $\mathcal{S}$ when the receiver is corrupted follow:

1. when the adversary $\mathcal{A}$ queries to $\mathsf{OTcrsGen}(1^\kappa)$ for a common reference string crs, the simulator $\mathcal{S}$ invokes the key generation algorithm DTGen of the Damgård-Jurik's double trapdoor encryption scheme to generate generate composite modulus of the form $N = pq$ that is a product of two safe primes p and q (i.e., $p = 2p' + 1$, $q = 2q' + 1$), a cyclic group $G \subseteq Z_N^*$ of order N' ($N' = p'q'$), and n random generators $g_1, \ldots, g_n$ of G and n random elements $(h_1, \ldots, h_n)$ such that $h_i = g_i^{x_i} \bmod N^2$, $x_i \in_U Z_{N'}$, $i = 1, \ldots, n$. The simulator keeps the auxiliary strings (p,q) and $(x_1, \ldots, x_n)$ secret. Let crs$_{DE} = <\mathrm{des}(G), (g_1, h_1), \ldots, (g_n, h_n), N >$.
   The simulator then invokes the key generation algorithms $(\mathsf{GenMC}, \mathsf{EKey})$ of the Damgård-Nielsen's mixed commitment scheme to generate a global public key $\widetilde{N}$ and n equivocable keys $(K_1, \ldots, K_n)$, where $K_j = \psi(0, k_j)$ (i.e., all K_j are E-keys). Let crs$_{MC} = < \widetilde{N}, K_1, \ldots, K_n >$, The auxiliary string is $(k_1, \ldots, k_n)$. The simulator keeps the auxiliary information $\widetilde{p}$ and $\widetilde{q}$ such that $\widetilde{N} = \widetilde{p}\widetilde{q}$, $\widetilde{p} = 2\widetilde{p}' + 1$, $\widetilde{q} = 2\widetilde{q}' + 1$ secret.
   Let crs $= ($crs$_{DE}$, crs$_{MC})$. $\mathcal{F}_{crs}^D$ returns (sid, crs) to the adversary $\mathcal{A}$;
2. The simulator $\mathcal{S}$ invokes the Damgård-Nielsen's mixed commitment scheme to generate a dummy database $(C_1, \ldots, C_n)$ for the retrieved messages.
3. Given crs and tpk, the simulator $\mathcal{S}$ checks the following three conditions: (1) checking $g \in Z_N^*$ and $h \in Z_N^*$; (2) checking $g^2 \bmod N \neq 1$ and $h^2 \bmod N \neq 1$. If any of two conditions is violated, then outputs $\perp$; otherwise, $\mathcal{S}$ extracts an index σ by testing the equation $h \stackrel{?}{=} g^{x_\sigma}$ ($i = 1, \ldots, n$). $\mathcal{S}$ sends σ to the ideal functionality $\mathcal{F}_{OT_1^n}$ and obtains m_σ.
4. $\mathcal{S}$ modifies the internal states to generate a transcript that is consistent with the previously generated ciphertexts $(C_1, \ldots, C_n)$ according to the following strategy: given C_σ (the encryption of dummy message with randomness r_σ), the simulator extracts a new randomness r'_σ from the equation $K_\sigma^0 r_\sigma^{\widetilde{N}} = K_\sigma^{m_\sigma} r_\sigma'^{\widetilde{N}}$, where $K_\sigma = \psi(0, k_\sigma)$.
   The simulator sends n ciphertexts $(E_1, \ldots, E_n)$ together with simulated transcripts of ZK-PKoE and ZK-PKoEQ to the adversary $\mathcal{A}$.

Let $\text{IDEAL}_{\mathcal{F},\mathcal{S},\mathcal{Z}}$ be the view of ideal world adversary $\mathcal{S}$ according to the description of simulation steps described above and $\text{REAL}_{\pi,\mathcal{A},\mathcal{Z}}$ be the view of real world adversary $\mathcal{A}$ according to the description of protocol π. The only difference between $\text{REAL}_{\pi,\mathcal{A},\mathcal{Z}}$ and $\text{IDEAL}_{\mathcal{F},\mathcal{S},\mathcal{Z}}$ is the generation of n commitments $(C_1, \ldots, C_n)$. By key indistinguishability assumption of the underlying mixed commitment scheme, we know that $\text{REAL}_{\pi,\mathcal{A},\mathcal{Z}} \approx \text{IDEAL}_{\mathcal{F},\mathcal{S},\mathcal{Z}}$.

SIMULATING TRIVIAL CASES. We now consider the following trivial cases: (1) both S and R are honest; and (2) both S and R are corrupted.

If both sender and receiver are honest, we define a simulator $\mathcal{S}$ below:

1. $\mathcal{S}$ internally runs the honest S on input (sid, $m_0 = 0 \in Z_N, \ldots, m_n = 0 \in Z_N$);
2. $\mathcal{S}$ internally runs the honest R on input (sid, $\sigma = i$), where $i \leftarrow_R [1, n]$;
3. $\mathcal{S}$ activates the protocol as specified when the corresponding dummy party is activated in the ideal execution, and delivering all messages between its internal R and S to $\mathcal{A}$.

Let $\text{IDEAL}_{\mathcal{F},\mathcal{S},\mathcal{Z}}$ be the view of ideal world adversary $\mathcal{S}$ according to the simulation steps described above and $\text{REAL}_{\pi,\mathcal{A},\mathcal{Z}}$ be the view of real world adversary $\mathcal{A}$ according to the description of protocol π. Since the underlying double-trapdoor cryptosystem is semantically secure assuming that the decisional Diffie-Hellman problem in G is hard, it follows that $\text{REAL}_{\pi,\mathcal{A},\mathcal{Z}} \approx \text{IDEAL}_{\mathcal{F},\mathcal{S},\mathcal{Z}}$.

When both S and R are corrupted by the real world adversary $\mathcal{A}$, the simulator generates what the corrupted parties generated (i.e., both the sender's messages and the receiver's messages which put together, forms the entire transcript). Thus, the simulator's task is trivial in this case.

Combining the above statements, we know that the proposed 1-out-of-n oblivious transfer protocol OT_1^n is universally composable in the $\mathcal{F}_{\text{crs}}^{\mathcal{D}}$-hybrid model assuming that the decisional Diffie-Hellman problem over G is hard.   □

5   Conclusion

In this paper, a construction of 1-out-of-n oblivious transfer protocol that is based on the Damgard and Juriks double trapdoor encryption scheme and the Damgard-Nielsens mixed commitment scheme has been constructed and analyzed. We have shown that the proposed oblivious transfer protocol realizes the universally composable security in the common reference model under the joint assumptions that the decisional Diffie-Hellman problem and the decisional composite residuosity problem are hard as well as all knowledge proof protocols applied are zero-knowledge in the presence of static adversaries.

References

1. Asharov, G., Lindell, Y., Schneider, T., et al.: More efficient oblivious transfer and extensions for faster secure computation. In: ACM Conference on Computer and Communications Security, pp. 535–548 (2013)
2. Brassard, G., Crépeau, C., Robert, J.M.: All-or-nothing disclosure of secrets. In: Odlyzko, A.M. (ed.) CRYPTO 1986. LNCS, vol. 263, pp. 234–238. Springer, Heidelberg (1987)
3. David, B., Dowsley, R., Nascimento, A.C.A.: Universally composable oblivious transfer based on a variant of LPN. In: Gritzalis, D., Kiayias, A., Askoxylakis, I. (eds.) CANS 2014. LNCS, vol. 8813, pp. 143–158. Springer, Heidelberg (2014)
4. Canetti, R.: A new paradigm for cryptographic protocols. In: FOCS, pp. 136–145 (2001)
5. Canetti, R.: Obtaining universally compoable security: towards the bare bones of trust. In: Kurosawa, K. (ed.) ASIACRYPT 2007. LNCS, vol. 4833, pp. 88–112. Springer, Heidelberg (2007)
6. Cramer, R., Damgård, I.B., Schoenmakers, B.: Proof of partial knowledge and simplified design of witness hiding protocols. In: Desmedt, Y.G. (ed.) CRYPTO 1994. LNCS, vol. 839, pp. 174–187. Springer, Heidelberg (1994)
7. Canetti, R., Lindell, Y., Ostrovsky, R., Sahai, A.: Universally composable two-party and multi-party secure computation. In: STOC, pp. 494–503 (2002)
8. Canetti, R., Fischlin, M.: Universally composable commitments. In: Kilian, J. (ed.) CRYPTO 2001. LNCS, vol. 2139, pp. 19–40. Springer, Heidelberg (2001)
9. Camenisch, J.L., Neven, G., Shelat, A.: Simulatable adaptive oblivious transfer. In: Naor, M. (ed.) EUROCRYPT 2007. LNCS, vol. 4515, pp. 573–590. Springer, Heidelberg (2007)
10. Choi, S.G., Hwang, K.-W., Katz, J., Malkin, T., Rubenstein, D.: Secure multiparty computation of boolean circuits with applications to privacy in on-line marketplaces. In: Dunkelman, O. (ed.) CT-RSA 2012. LNCS, vol. 7178, pp. 416–432. Springer, Heidelberg (2012)
11. Cramer, R., Shoup, V.: Universal hash proofs and a paradigm for adaptive chosen ciphertext secure public-key encryption. In: Knudsen, L.R. (ed.) EUROCRYPT 2002. LNCS, vol. 2332, p. 45. Springer, Heidelberg (2002)
12. Crépeau, C.: Equivalence between two flavours of oblivious transfers. In: Pomerance, C. (ed.) CRYPTO 1987. LNCS, vol. 293, pp. 350–354. Springer, Heidelberg (1988)
13. Damgård, I.B., Nielsen, J.B.: Perfect hiding and perfect binding universally composable commitment schemes with constant expansion factor. In: Yung, M. (ed.) CRYPTO 2002. LNCS, vol. 2442, pp. 581–596. Springer, Heidelberg (2002)
14. Dubovitskaya, M.: Cryptographic protocols for privacy-preserving access control in databases. Doctoral dissertation, Diss., Eidgenosische Technische Hochschule ETH Zurich, no. 21835 (2014)
15. Damgård, I., Jurik, M.: A length-flexible threshold cryptosystem with applications. In: ACISP, pp. 350–364 (2003)
16. Even, S., Goldreich, O., Lempel, A.: A randomized protocol for signing contracts. Commun. ACM **28**(6), 637–647 (1985)
17. Guleria, V., Dutta, R.: Universally composable issuer-free adaptive oblivious transfer with access policy. Secur. Commun. Netw. 8, 3615–3633 (2015)

18. Asharov, G., Lindell, Y., Schneider, T., Zohner, M.: More efficient oblivious transfer extensions with security for malicious adversaries. In: Oswald, E., Fischlin, M. (eds.) EUROCRYPT 2015. LNCS, vol. 9056, pp. 673–701. Springer, Heidelberg (2015)
19. Goldreich, O., Micali, S., Wigderson, A.: How to play any mental game-or-a completeness theorem for protocols with honest majority. In: STOC, pp. 218–229 (1987)
20. Green, M., Hohenberger, S.: Blind identity-based encryption and simulatable oblivious transfer. In: ASIACRYPT, pp. 265–282 (2007)
21. Green, M., Hohenberger, S.: Universally composable adaptive oblivious transfer. In: ASIACRYPT (2008)
22. Huang, Y., Evans, D., Katz, J., Malka, L.: Faster secure two-party computation using garbled circuits. In: 20th USENIX Security Symposium, San Francisco, CA, 8–12 August 2011
23. Kilian, J.: Founding cryptography on oblivious transfer. In: STOC, pp. 20–31 (1988)
24. Lindell, Y., Pinkas, B.: An efficient protocol for secure two-party computation in the presence of malicious adversaries. Commun. ACM **28**(2), 312–350 (2015)
25. Manoj, K., Praveen, I.: A fully simulatable oblivious transfer scheme using vector decomposition. In: Jain, L.C., Patnaik, S., Ichalkaranje, N. (eds.) Intelligent Computing, Communication and Devices, pp. 131–137. Springer, India (2015)
26. Naor, M., Pinkas, B.: Efficient oblivious transfer protocols. In: SODA, pp. 448–457 (2001)
27. Naor, M., Pinkas, B.: Oblivious transfer with adaptive queries. In: Wiener, M. (ed.) CRYPTO 1999. LNCS, vol. 1666, pp. 573–590. Springer, Heidelberg (1999)
28. Peikert, C., Vaikuntanathan, V., Waters, B.: A framework for efficient and composable oblivious transfer. In: Wagner, D. (ed.) CRYPTO 2008. LNCS, vol. 5157, pp. 554–571. Springer, Heidelberg (2008)
29. Michael O.Rabin.: How to exchange secrets by oblivious transfer. Technical report TR-81, Aiken Computation Laboratory, Harvard University (1981)
30. Kalai, Y.T.: Smooth projective hashing and two-message oblivious transfer. In: Cramer, R. (ed.) EUROCRYPT 2005. LNCS, vol. 3494, pp. 78–95. Springer, Heidelberg (2005)
31. Andrew Chi-Chih Yao.: Protocols for secure computations (extended abstract). In: FOCS, pp. 160–164 (1982)
32. Zhu, H.: Round optimal universally composable oblivious transfer protocols. In: Baek, J., Bao, F., Chen, K., Lai, X. (eds.) ProvSec 2008. LNCS, vol. 5324, pp. 328–334. Springer, Heidelberg (2008)
33. Zhu, H., Bao, F.: Adaptive and composable oblivious transfer protocols (short paper). In: Qing, S., Mitchell, C.J., Wang, G. (eds.) ICICS 2009. LNCS, vol. 5927, pp. 483–492. Springer, Heidelberg (2009)

Public Key Encryption with Distributed Keyword Search

Veronika Kuchta$^{(\boxtimes)}$ and Mark Manulis

Department of Computing, University of Surrey, Guildford, UK
v.kuchta@surrey.ac.uk, mark@manulis.eu

Abstract. In this paper we introduce Threshold Public Key Encryption with Keyword Search (TPEKS), a variant of PEKS where the search procedure for encrypted keywords is distributed across multiple servers in a threshold manner. TPEKS schemes offer stronger privacy protection for keywords in comparison to traditional PEKS schemes. In particularly, they prevent *keyword guessing attacks* by malicious servers. This protection is not achievable in a single-server PEKS setting.

We show how TPEKS can be built generically from any anonymous Identity-Based Threshold Decryption (IBTD), assuming the latter is indistinguishable, anonymous and robust. In order to instantiate our TPEKS construction we describe an efficient IBTD variant of the Boneh-Franklin IBE scheme. We provide an appropriate security model for such IBTD schemes and give an efficient construction in the random oracle model.

TPEKS constructions are particularly useful in distributed cloud storage systems where none of the servers alone is sufficiently trusted to perform the search procedure and where there is a need to split this functionality across multiple servers to enhance security and reliability.

1 Introduction

Cloud computing provides convenient, on-demand network access to shared services and applications over the Internet. The main advantages of cloud computing are the virtually unlimited data storage capabilities, universal data access, and savings on hardware and software expenses. Despite the many technical and economical advantages, availability and data privacy are amongst those issues that prevent potential users from trusting the cloud services. The reason of these concerns is that upon outsourcing their data to the cloud, users lose control over their data.

While for better availability it is advisable to distribute copies of data across multiple cloud servers, for data privacy and its protection from unauthorized access the use of complete encryption prior to outsourcing is indispensable. If the data is encrypted and the user wishes to access certain files at a later stage, the cloud needs to perform the search and retrieve the corresponding ciphertexts. In order to facilitate the search process each outsourced file is typically associated with a set of keywords. Since adding plaintext keywords [11] to each file prior to

© Springer International Publishing Switzerland 2016
M. Yung et al. (Eds.): INTRUST 2015, LNCS 9565, pp. 62–83, 2016.
DOI: 10.1007/978-3-319-31550-8_5

outsourcing would leak information about the file contents to the cloud services, a better solution is to encrypt the associated keywords and provide cloud services with the ability to search for encrypted keywords. This permission comes in form of trapdoors allowing cloud services to test whether an encrypted file contains keywords for which the trapdoors were derived by the user. Such searchable encryption techniques are particularly helpful to protect outsourcing of sensitive files, e.g., those containing medical and health records [6,20,21].

A range of encryption schemes supporting keyword search have been proposed, based on symmetric encryption (e.g., [12,14,16,27]) and public-key encryption (e.g., [7,9,17,18]) techniques. While some schemes can cope only with single keywords (e.g., [26,29]), which is too restrictive in practice, more advanced schemes (e.g., [11,22,25,28]) can process multiple keywords. The majority of searchable encryption schemes issue trapdoors for keywords and any party in possession of those trapdoors can perform the search procedure on its own. In a cloud-based storage setting this imposes a single point of trust with regard to the search functionality. When it comes to the use of multiple cloud services for better availability, a distributed search approach would therefore help to reduce this trust requirement. Encryption schemes supporting distributed keyword search procedures in a cloud environment exist so far only in the symmetric setting, namely in [30], those constructions however were not formally modeled and analyzed.

Our Threshold Public Key Encryption with Keyword Search (TPEKS).
We model security and propose first constructions of Threshold Public Key Encryption with Keyword Search (TPEKS), where the ability to search over encrypted keywords requires participation of at least t parties, each equipped with its own trapdoor share. Main benefits of TPEKS over traditional PEKS constructions include the distribution of trust across multiple servers involved in a search procedure and more importantly stronger privacy protection of keywords against keyword guessing attacks [10,24], based on which information about keywords can be revealed from associated trapdoors; notably, all single-server-based PEKS constructions are inherently vulnerable to keyword guessing attacks.

Our security model for TPEKS is motivated by the security goals and known attacks on single-server-based PEKS constructions (e.g. [1,3,4,10,13,19]). The concept of PEKS was introduced by Boneh et al. [7], along with the formalization of two security goals: indistinguishability and consistency. While indistinguishability aims at privacy of encrypted keywords, consistency aims to prevent false positives, where a PEKS ciphertext created for one keyword can successfully be tested with a trapdoor produced for another keyword. Initially, PEKS constructions were able to search for individual keywords, whereas later schemes were designed to handle conjunctive queries on encrypted keywords [17,18,23] and thus process keyword sets within a single search operation. The majority of PEKS schemes offer security against chosen-plaintext attacks [1,7,10,13] and only few constructions remain secure against chosen-ciphertext attacks, e.g. [15]. The vulnerability of PEKS constructions, e.g. [7], against (offline) keyword guessing attacks was discovered by Byun et al. [10]. In short, a keyword guessing attack can be mounted by creating a PEKS ciphertext for some candidate keyword and then testing this ciphertext with the given trapdoor. Obviously, this attack works if keywords have low entropy, which is what typically happens in practice. As shown by Jeong et al. [19], keyword guessing attacks are inherent to all single-server based PEKS constructions with consistency, a necessary security property of PEKS. Through secret sharing of the trapdoor information across multiple servers, TPEKS significantly reduces the risk of keyword guessing attacks, which are modeled as part of the indistinguishability property.

In the design of our TPEKS construction we extend the ideas underlying the transformation by Abdalla et al. [1] for building indistinguishable and computationally consistent (single-server) PEKS from anonymous Identity-Based Encryption (IBE). Although our transformation also treats identities as keywords, it assumes a different building block, namely anonymous Identity-Based Threshold Decryption (IBTD), which extends IBE by the distributed decryption process for which a threshold number t-out-of-n servers contribute with their own decryption shares. We show that while IBTD anonymity is essential for the indistinguishability (with resistance to keyword guessing attacks) of the constructed TPEKS scheme, IBTD indistinguishability informs computational consistency property of the TPEKS scheme. Aiming to instantiate our TPEKS construction, we propose an anonymous IBTD scheme, as a modification of the well-known anonymous IBE scheme by Boneh and Franklin (BF) [8]. This modification is performed by distributing the decryption process of the original BF-scheme.

2   Anonymous Identity-Based Threshold Decryption

We start with the definitions of Identity-Based Threshold Decryption (IBTD) along with its security properties: indistinguishability, anonymity and robustness. Our IBTD model extends the model from [5], where this primitive along with the indistinguishability property was introduced, by additional anonymity requirement, which also requires some small modifications to the assumed syntax of the IBTD decryption process in comparison to [5].

2.1   IBTD Syntax and Security Goals

We formalize the IBTD syntax in Definition 1. In contrast to [5], we treat the validity checking process for decryption shares implicitly as part of the decryption algorithm Dec, whereas in [5] this property was outsourced into a separate verification algorithm, aiming at public verifiability of individual decryption shares. In our case, where we additionally require the IBTD scheme to be anonymous such syntax change is necessary, as discussed in Remark 1.

Definition 1 (Identity-Based Threshold Decryption (IBTD)). *An IBTD scheme consists of the following algorithms* (Setup, KeyDer, Enc, ShareDec, Dec)*:*

Setup$(n, t, 1^k)$: *On input the number of decryption servers n, a threshold para-meter t, $(1 \leq t \leq n)$ and a security parameter 1^k, it outputs a master public key mpk and a master secret key msk.*

KeyDer(mpk, msk, id, t, n) : *On input a master public key mpk, master secret key msk, identity id, and threshold parameters t, n, it computes the secret key sk_{id} for identity id and outputs the private tuple $(i, sk_{id,i})$ for server S_i $1 \leq i \leq n$.*

Enc(mpk, id, m) : *On input mpk, id, m it outputs a ciphertext C.*

ShareDec$(mpk, (i, sk_{id,i}), C)$: *On input a master public key mpk, secret shares (i, sk_{id_i}) for servers $1 \leq i \leq n$ and ciphertext C. It outputs decryption shares δ_i for $1 \leq i \leq n$.*

Dec$(mpk, \{\delta_i\}_{i \in \Omega}, C)$: *On input a master public key mpk, a set of decryption shares $\{\delta_i\}_{i \in \Omega}$, where $|\Omega| \geq t$ and a ciphertext C. It outputs m (which can also be $\perp$ to indicate a failure).*

The following Definition 2 formalizes IBTD indistinguishability against chosen-ciphertext attacks (IBTD-IND-CCA) and bears similarities with the corresponding definition for IBE [8]; namely, our experiment takes into account the threshold nature of the decryption algorithm allowing the adversary to reveal up to $t - 1$ secret key shares.

Definition 2 (IBTD Indistinguishability). *Let $\mathcal{A}_{ind}$ be a PPT adversary against the IBTD-IND-CCA security of the IBTD scheme, associated with the following experiment* $\mathbf{Exp}_{\mathcal{A}_{ind}}^{\text{IBTD-IND-CCA}-b}(1^\lambda)$:

1. $(mpk, msk) \xleftarrow{r} \text{Setup}(1^\lambda, t, n)$
2. *Let* **List** *be a list comprising* (id, S_{id})*, where* $S_{id} := \{(1, sk_{id,1}), \ldots, (n, sk_{id,n})\}$ *and* $(i, sk_{id,i})$ *are the outputs of* KeyDer(mpk, msk, id, t, n) *algorithm. Note: at the beginning of the experiment the list is empty.*
3. $(id^*, m_0, m_1, state) \xleftarrow{r} \mathcal{A}_{ind}^{\mathcal{O}\text{KeyDer}(\cdot), \mathcal{O}\text{Dec}(\cdot)}(find, mpk)$
4. *if* $(id^*, S_{id^*}) \notin$ **List***, run* $(sk_{id^*}, (1, sk_{id^*,1}), \ldots, (n, sk_{id^*,n})) \xleftarrow{r}$ KeyDer (mpk, msk, id^*, t, n), *set* $S_{id^*} := \{(1, sk_{id^*,1}), \ldots, (n, sk_{id^*,n})\}$, *add* (id^*, S_{id^*}) *to* **List**
5. *pick* $b \in \{0, 1\}$*, compute* $C^* \xleftarrow{r} \text{Enc}(mpk, id^*, m_b)$
6. $b' \xleftarrow{r} \mathcal{A}_{ind}^{\mathcal{O}\text{KeyDer}(\cdot), \mathcal{O}\text{Dec}(\cdot)}(guess, C^*, mpk)$

The experiment outputs 1 if all of the following holds:

– $b' = b$
– $\mathcal{A}$ *issued at most* $t - 1$ *queries* $\mathcal{O}\text{KeyDer}(id^*, i)$
– $\mathcal{A}_{ind}$ *did not query* $\mathcal{O}\text{Dec}(id^*, C^*)$

where the two oracles are defined as follows:

$\mathcal{O}\text{KeyDer}(id, i)$: *On input* (id, i) *check whether* $(id, S_{id}) \in$ **List***. If so, parse* S_{id} *as* $\{(1, sk_{id,1}), \ldots, (n, sk_{id,n})\}$ *and output* (i, sk_{id_i})*. If* $(id, S_{id}) \notin$ **List** *run* $S \xleftarrow{r}$ KeyDer(mpk, msk, id, t, n)*. Add* (id, S_{id}) *to* **List***, output* $(i, sk_{id,i})$*.*

$\mathcal{O}Dec(id, C)$: On input (id, C) check whether $(id, S_{id}) \in$ **List**. If so, parse S_{id} as $\{(1, sk_{id,1}), \ldots, (n, sk_{id,n})\}$, run $\delta_i \xleftarrow{r} \mathtt{ShareDec}(mpk, (i, sk_{id,i}), C)$ for $i \in [n]$. Take at least t-out-of-n decryption shares δ_i, run $\mathtt{Dec}(\{\delta_i\}_{i \in \Omega}, C)$, where $|\Omega| \geq t$ and output m or 0. If $(id, S_{id}) \notin$ **List**, compute $S_{id} \xleftarrow{r} \mathtt{KeyDer}(mpk, msk, id, t, n)$ add (id, S_{id}) to the **List**. Compute $\delta_i \xleftarrow{r} \mathtt{ShareDec}(mpk, (i, sk_{id,i}), C)$, where $i \in [n]$. Take at least t-out-of-n decryption shares δ_i, run $\mathtt{Dec}(\{\delta_i\}_{i \in \Omega}, C)$, output m or 0.

$\mathcal{A}_{ind}$'s success is given as

$$\mathbf{Adv}_{\mathcal{A}_{ind}}^{\text{IBTD-IND-CCA}-b}(1^\lambda) = \left| Pr\left[\mathbf{Exp}_{\mathcal{A}_{ind}}^{\text{IBTD-IND-CCA}-1}(1^\lambda) = 1\right]\right.$$
$$\left. -Pr\left[\mathbf{Exp}_{\mathcal{A}_{ind}}^{\text{IBTD-IND-CCA}-0}(1^\lambda) = 1\right]\right|.$$

The scheme is IBTD-IND-CCA secure if $\mathbf{Adv}_{\mathcal{A}_{ind}}^{\text{IBTD-IND-CCA}-b}$ is negligible.

In Definition 3 we model anonymity as a new property for IBTD schemes. Our definition bears some similarity with the anonymity property for IBE schemes as defined, e.g. in [1], except that we consider chosen-ciphertext attacks through the inclusion of the decryption oracle and account for the threshold setting by allowing the anonymity adversary to reveal up to $t - 1$ secret key shares. This latter ability is particularly important for achieving protection against keyword guessing attacks for our transformation from IBTD to TPEKS.

Definition 3 (IBTD Anonymity). Let $\mathcal{A}_{ano}$ be a probabilistic polynomial-time adversary against the IBTD-ANO-CCA security of the IBTD scheme, associated with the following experiment $\mathbf{Exp}_{\mathcal{A}_{ano}}^{\text{IBTD-ANO-CCA}-b}(1^\lambda)$

1. $(mpk, msk) \xleftarrow{r} \mathtt{Setup}(1^\lambda, t, n)$
2. Let **List** be a list storing (id, S_{id}), where $S_{id} := \{(1, sk_{id,1}), \ldots, (n, sk_{id,n})\}$ and $(i, sk_{id,i})$ are the outputs of $\mathtt{KeyDer}(mpk, msk, id, t, n)$ algorithm. Note: at the beginning of the experiment the list is empty.
3. $(id_0, id_1, m^*, state) \xleftarrow{r} A_{ano}^{\mathcal{O}\mathtt{KeyDer}(\cdot), \mathcal{O}\mathtt{Dec}(\cdot)}(find, mpk)$
4. if $(id_0, S_{id_0}) \notin$ **List**: $(sk_{id_0}, (1, sk_{id_0,1}), \ldots, (n, sk_{id_0,n})) \xleftarrow{r} \mathtt{KeyDer}(mpk, msk, id_0, t, n)$, set $S_{id_0} := \{(1, sk_{id_0,1}), \ldots, (n, sk_{id_0,n})\}$, add (id_0, S_{id_0}) to **List**
5. if $(id_1, S_{id_1}) \notin$ **List**: $(sk_{id_1}, (1, sk_{id_1,1}), \ldots, (n, sk_{id_1,n})) \xleftarrow{r} \mathtt{KeyDer}(mpk, msk, id_1, t, n)$, set $S_{id_1} := \{(1, sk_{id_1,1}), \ldots, (n, sk_{id_1,n})\}$, add (id_1, S_{id_1}) to **List**
6. $b \in \{0, 1\}$; $C^* \xleftarrow{r} \mathtt{Enc}(mpk, id_b, m^*)$
7. $b' \xleftarrow{r} \mathcal{A}_{ano}^{\mathcal{O}\mathtt{KeyDer}(\cdot), \mathcal{O}\mathtt{Dec}(\cdot)}(guess, C^*, mpk)$.

The experiment outputs 1 if all of the following holds

- $b' = b$
- $\mathcal{A}_{ano}$ issued at most $t - 1$ queries $\mathcal{O}\mathtt{KeyDer}(id_0, i)$ and at most $t - 1$ queries $\mathcal{O}\mathtt{KeyDer}(id_1, i)$
- $\mathcal{A}_{ano}$ did not query $\mathcal{O}\mathtt{Dec}(id_0, C^*)$ or $\mathcal{O}\mathtt{Dec}(id_1, C^*)$

where the two oracles are defined as follows:

$\mathcal{O}$KeyDer(id, i) : *On input* (id, i) *check whether* $(id, S_{id}) \in$ **List**. *If so, parse* S_{id} *as* $\{(1, sk_{id,1}), \ldots, (n, sk_{id,n})\}$ *and output* (i, sk_{id_i}). *If* $(id, S_{id}) \notin$ **List** *run* $S_{id} \xleftarrow{r}$ KeyDer(mpk, msk, id, t, n). *Add* (id, S_{id}) *to* **List***, output* $(i, sk_{id,i})$.

$\mathcal{O}$Dec(id, C): *On input* (id, C) *check whether* $(id, S_{id} \in$ **List**. *If so, parse* S_{id} *as* $\{(1, sk_{id,1}), \ldots, (n, sk_{id,n})\}$, *compute* $\delta_i \xleftarrow{r}$ ShareDec$(mpk, (i, sk_{id,i}), C)$ *for* $i \in [n]$. *Take at least* t-*out-of-*n *decryption shares* δ_i, *run* Dec$(\{\delta_i\}_{i \in \Omega}, C)$, *where* $|\Omega| \geq t$ *and output* m *or* 0. *If* $(id, S_{id}) \notin$ **List***, compute* $S_{id} \xleftarrow{r}$ KeyDer(mpk, msk, id, t, n) *add* (id, S_{id}) *to the* **List***. Compute* $\delta_i \xleftarrow{r}$ ShareDec$(mpk, (i, sk_{id,i}), C)$, *where* $i \in [n]$. *Take at least* t-*out-of-*n *decryption shares* δ_i, *run* Dec$(\{\delta_i\}_{i \in \Omega}, C)$, *output* m *or* 0.

The advantage of $\mathcal{A}_{ano}$ *is defined as*

$$\mathbf{Adv}_{\mathcal{A}_{ano}}^{\text{IBTD-ANO-CCA}-b}(1^\lambda) = \left| Pr\left[\mathbf{Exp}_{\mathcal{A}_{ano}}^{\text{IBTD-ANO-CCA}-1}(1^\lambda) = 1 \right] \right.$$
$$\left. -Pr\left[\mathbf{Exp}_{\mathcal{A}_{ano}}^{\text{IBTD-ANO-CCA}-0}(1^\lambda) = 1 \right] \right|.$$

The scheme is IBTD-ANO-CCA *secure if* $\mathbf{Adv}_{\mathcal{A}_{ano}}^{\text{IBTD-ANO-CCA}}$ *is negligible.*

Remark 1. Without the aforementioned change to the IBTD syntax of the decryption process, in comparison to [5], we would not be able to allow adversarial access to up to $t - 1$ decryption shares for the challenge ciphertext in the above anonymity experiment. The ability to publicly verify individual decryption shares using the challenge ciphertext and a candidate identity (as in [5]) would rule out any meaningful definition of anonymity; in particular, a single decryption share for the challenge ciphertext would suffice to break its anonymity property. In fact, it can be easily verified that the IBTD construction in [5] is not anonymous according to our definition.

Robustness of IBTD. In the following definition we formalize IBTD robustness, meaning that the decryption algorithm will output $\perp$ with overwhelming probability if an IBTD ciphertext computed for some id is decrypted using $sk_{id'}$ for $id' \neq id$. Our definition of strong robustness extends the one for IBE schemes in [2] to the threshold decryption setting.

Definition 4 (IBTD-SROB-CCA). *Let* $\mathcal{A}_{rob}$ *be a probabilistic polynomial-time adversary against the* IBTD-SROB-CCA *security of the IBTD scheme, associated with the following experiment* $\mathbf{Exp}_{\mathcal{A}_{rob}}^{\text{IBTD-SROB-CCA}}(1^\lambda)$:

1. $(mpk, msk) \xleftarrow{r}$ Setup$(1^\lambda, t, n), b \xleftarrow{r} \{0,1\}$, **List** $:= \emptyset, I = \emptyset$
2. *Let* **List** *be a list storing* (id, S_{id_b}, I_b), *with* $S_{id_b} := \{(1, sk_{id_b, 1}), \ldots, (n, sk_{id_b, n})\}$ *and* $b \xleftarrow{r} \{0,1\}$, *where* $(i, sk_{id_b, i}) \leftarrow$ KeyDer(mpk, msk, id_b, t, n)
3. $(id_0^*, id_1^*, C^*, state) \xleftarrow{r} A_{rob}^{\mathcal{O}\text{KeyDer}(\cdot), \mathcal{O}\text{Dec}(\cdot)}(find, mpk)$
4. (i) *If* $id_0 = id_1$ *then return* 0.
   (ii) *If* $(id_0, S_{id_0}, I_0) \notin$ **List** *or* $(id_1, S_{id_1}, I_1) \notin$ **List***, return* 0.

(iii) If $|I_0| \geq t$ or $|I_1| \geq t$, then return 0. Else compute decryption shares $\delta_{0,i} \xleftarrow{r} \mathtt{ShareDec}(mpk, (i, sk_{id_0,i}), C)$, $m_0 \xleftarrow{r} \mathtt{Dec}(mpk, \{\delta_{0,i}\}_{i \in \Omega}, C)$ and $\delta_{1,i} \xleftarrow{r} \mathtt{ShareDec}(mpk, (i, sk_{id_1,i}), C)$, $m_1 \xleftarrow{r} \mathtt{Dec}(mpk, \{\delta_{1,i}\}_{i \in \Omega}, C)$. If $m_0 \neq \bot$ and $m_1 \neq \bot$ return 1.

$\mathcal{O}\mathtt{KeyDer}(id, i)$: On input (id, i) check whether $(id, S_{id}) \notin$ **List**. If so, compute $S \xleftarrow{r} \mathtt{KeyDer}(mpk, msk, id, t, n)$, where $S_{id} := \{(1, sk_{id,1}), \ldots, (n, sk_{id,n})\}$ and $I \subset [1, n]$, add (id, S_{id}, I) to **List**. Then add i to I and return $(i, sk_{id,i})$.

$\mathcal{O}\mathtt{Dec}(id, C)$: On input (id, C) check whether $(id, S_{id}) \notin$ **List**. If so, compute $S_{id} \xleftarrow{r} \mathtt{KeyDer}(mpk, msk, id, t, n)$ add (id, S_{id}, I) to **List**. Finally compute $\delta_i \xleftarrow{r} \mathtt{ShareDec}(mpk, (i, sk_{id,i}), C)$, $i \in [n]$, $m \leftarrow \mathtt{Dec}(mpk, \{\delta_i\}_{i \in \Omega}, C)$, where $|\Omega| \geq t$. Output m.

We have: $\mathbf{Adv}_{\mathcal{A}_{rob}}^{\text{IBTD-SROB-CCA}}(1^\lambda) = \left| Pr \left[\mathbf{Exp}_{\mathcal{A}_{rob}}^{\text{IBTD-SROB-CCA}}(1^\lambda) = 1 \right] \right|$.

2.2   An Anonymous IBTD Scheme Based on Boneh-Franklin IBE

We propose a concrete IBTD construction, based on Boneh-Franklin IBE [8] where we apply secret sharing to individual private keys and to the decryption procedure. In particular, upon receiving at least t decryption shares, the decryption algorithm outputs either the message m or 0 (to indicate a failure). In contrast to the so-far only IBTD scheme in [5], which also builds on the Boneh-Franklin IBE, our construction is anonymous. Abdalla et al. [2] proved that Boneh-Franklin IBE is robust in the random oracle model.

Definition 5 (Anonymous IBTD Scheme). $\mathtt{Setup}(n, t, 1^k)$: On input a security parameter 1^k, it specifies $\mathbb{G}, \mathbb{G}_T$ of order $q \geq 2^k$, chooses a generator $g \in \mathbb{G}$, specifies a bilinear map $e : \mathbb{G} \times \mathbb{G} \to \mathbb{G}_T$, random oracles H_1, H_2, H_3, H_4 s.t. $H_1 : \{0, 1\}^* \to \mathbb{G}$; $H_2 : \mathbb{G}_T \to \{0, 1\}^\ell$; $H_3 : \{0, 1\}^\ell \times \{0, 1\}^\ell \to \mathbb{Z}_q^*$; $H_4 : \{0, 1\}^\ell \to \{0, 1\}^\ell$. The message space is $\mathcal{M} = \{0, 1\}^\ell$. The ciphertext space is $\mathcal{C} = \mathbb{G}^* \times \{0, 1\}^\ell$. It picks $x \xleftarrow{r} \mathbb{Z}_q^*$ and computes $Y = g^x$. It returns $mpk = (\mathbb{G}, \mathbb{G}_T, q, g, e, H_1, H_2, H_3, H_4, Y)$ and $msk = x$.

$\mathtt{KeyDer}(mpk, msk, id, t, n)$: On input an identity id, computes $Q_{id} = H_1(id) \in \mathbb{G}^*$ where $1 \leq t \leq n < q$ and using $msk = x$ it computes $sk_{id} = Q_{id}^x = H_1(id)^x$. It picks $a_1, \ldots, a_{t-1} \xleftarrow{r} \mathbb{G}$, computes a polynomial $f(u) = f(0) + \sum_{i=1}^{t-1} a_i u^i$, where $f(u) \in \mathbb{Z}_q(u)$, $u \in \mathbb{N} \cup \{0\}$, s.t. $f(0) = x$. It outputs n master key shares $(i, sk_{id,i})$, where $sk_{id,i} = Q_{id}^{f(i)}$ for $i \in \{1, \ldots, n\}$. To derive the private key let $\lambda_1, \ldots, \lambda_t \in \mathbb{Z}_q$ be the Lagrange coefficients, s.t. $x = \sum_{i=0}^{t-1} \lambda_i f(i)$.

$\mathtt{Enc}(mpk, id, m)$: On input the public key mpk, a plaintext $m \in \{0, 1\}^\ell$ and an identity $id \in \{0, 1\}^*$ computes $Q_{id} = H_1(id) \in \mathbb{G}^*$, chooses $\sigma \xleftarrow{r} \{0, 1\}^\ell$, sets $r = H_3(\sigma, m)$. It computes $U = g^r, V = \sigma \oplus H_2(\kappa_{id}), W = m \oplus H_4(\sigma)$, where $\kappa_{id} = e(Q_{id}, Y)^r$ and returns $C = \langle U, V, W \rangle$.

$\mathtt{ShareDec}(mpk, (i, sk_{id,i}), C)$: On input a ciphertext $C = \langle U, V, W \rangle$ and a secret key share $(i, sk_{id,i})$, the algorithm returns $\delta_i = e(sk_{id,i}, U) = e(Q_{id}^{f(i)}, U)$.

$Dec(mpk, \{\delta_i\}_{i \in \Omega}, C)$: *Given a set of decryption shares* $\{\delta_i\}_{i \in \Omega}$, $|\Omega| \geq t$ *and a ciphertext* $C = \langle U, V, W \rangle$, *it computes Lagrange coefficients* $\lambda_i = \prod_{j \in \Omega, j \neq i}^{t} \frac{-j}{i-j}$ *and reconstructs* $\kappa_{id} := \prod_{i=1}^{t} \delta_i^{\lambda_i}$. *It computes* $\sigma = V \oplus H_2(e(\kappa_{id}))$, $m = W \oplus H_4(\sigma)$, *and* $r = H_3(\sigma, m)$. *Then, if* $U = g^r$ *it outputs* m; *otherwise it outputs 0. (Note that the equality check* $U = g^r$ *is essential for detecting inconsistent ciphertexts and by this preventing chosen ciphertext attacks.)*

Correctness. The proposed IBTD scheme is correct since κ_{id} computed by the decryption algorithm is the same as was used by the encryption algorithm, i.e. $\kappa_{id} = \prod_{i=1}^{t} e(sk_{id,i}, U)^{\lambda_i} = \prod_{i=1}^{t} e(sk_{id,i}^{\lambda_i}, g^r) = e\left(\prod_{i=1}^{t} Q_{id}^{f(i)\lambda_i}, g\right)^r = e\left(Q_{id}^{\sum_{i=0}^{t-1} f(i)\lambda_i}, g^r\right) = e(Q_{id}^x, g^r) = e(Q_{id}, Y)^r$.

2.3 Security Analysis

The overall security of our IBTD scheme relies on the well-known Decisional Bilinear Diffie-Hellman (DBDH) assumption and the random oracle model.

Definition 6 (DBDH Assumption). *The Decisional Bilinear Diffie-Hellman (DBDH) assumption in the bilinear map setting* $(q, \mathbb{G}, \mathbb{G}_T, e, g)$, *where* $e : \mathbb{G} \times \mathbb{G} \rightarrow \mathbb{G}_T$ *and* g *is the generator of* $\mathbb{G}$ *states that for any PPT distinguisher* $\mathcal{A}$ *the following advantage is negligible:* $\mathbf{Adv}_{\mathcal{A}}^{DBDH} = \left| Pr \left[\mathcal{A}(g, g^a, g^b, g^c, e(g,g)^{abc}) = 1 \right] - Pr \left[\mathcal{A}(g, g^a, g^b, g^c, e(g,g)^z) = 1 \right] \right|$, *where the probability is taken over the random choices of* $a, b, c, z \in \mathbb{Z}_q$ *and the random bits of* $\mathcal{A}$.

Theorem 1 (IBTD-IND-CCA). *Our IBTD scheme from Definition 5 is IBTD-IND-CCA secure under the DBDH assumption in the random oracle model.*

Proof. For the proof of this theorem we refer to Appendix A.

Theorem 2 (IBTD-ANO-CCA). *Our IBTD scheme from Definition 5 is IBTD-ANO-CCA secure under the DBDH assumption in the random oracle model.*

Proof. Since the proof of anonymity is similar to the proof of indistinguishability we only give a sketch. The simulator sets $Q_{id_\beta} = (g^b)^{\nu_\beta}$, for a random $\beta \in \{0,1\}$ and $Y = g^c$. It responds to the key derivation and decryption queries like in the indistinguishability proof. Upon finishing phase 1, $\mathcal{A}_{ano}$ outputs a message m and two identities id_0, id_1 it wants to be challenged on. The challenge ciphertext is computed as follows. $\mathcal{A}_{ano}$ sets $Q_{id_\beta} = (g^b)^{\nu_\beta}$ and $Y = g^c$ such that $e(Q_{id_\beta}, Y) = e(g^b, g^c)$ and $\kappa_{id} = Z$, where Z is the value from BDH instance. It chooses $s \in \{0,1\}^\ell$ uniformly at random. $\mathcal{A}_{BDH}$ gives $C^* = (g^a, s \oplus H_2(Z), m \oplus H_4(s)), \beta \in \{0,1\}$ to $\mathcal{A}_{ano}$. U is chosen uniformly at random by the encryption algorithm. V depends on the randomly chosen $s \in \{0,1\}^\ell$ and $H_2(Z)$. Since Z is randomly chosen, it does not depend on id_β. Also W is independent of id_β and therefore the ciphertext has the same distribution for both $\beta \in \{0,1\}$ and $\mathcal{A}_{ano}$ will have 0 advantage to distinguish between id_0 and id_1.

Theorem 3 (IBTD-SROB-CCA). *Our IBTD scheme from* Definition 5 *is unconditionally IBTD-SROB-CCA secure in the random oracle model.*

Proof. (Sketch) We show that IBTD-SROB-CCA property holds in the random oracle model. Assume $\mathcal{A}_{rob}$ be a IBTD-SROB-CCA adversary that is given the master key x. He can receive at most $t - 1$ secret shares from $\mathcal{O}\mathsf{KeyDer}$. We note, that H_1 is a map to $\mathbb{G}^*$, where all the outputs of this map are elements of order q. That means that the probability of finding two different identities $id_1 \neq id_2$ such that $H_1(id_1) = H_1(id_2)$ is negligible. Since $Y = g^x \in \mathbb{G}^*$ we have that κ_{id_1} and κ_{id_2} are not equal and of order q. Since H_3 maps into $\mathbb{Z}_q^*$, then $\kappa_{id_1}^r$ and $\kappa_{id_2}^r$ are different. Assuming H_2 as a random oracle, means that $H_2(\kappa_{id_1}^r) \neq H_2(\kappa_{id_2}^r)$. Decryption under different identities yields therefore two different values $\sigma_1 \neq \sigma_2$. In order for the ciphertext to be valid for both id's it should hold that $r = H_3(\sigma_1, m_1) = H_3(\sigma_2, m_2)$, which happens with negligible probability. It follows that our IBTD scheme is IBTD-SROB-CCA secure.

3   Threshold Public Key Encryption with Keyword Search

We start by defining the TPEKS syntax and its security goals. Towards the end of this section we propose a general transformation for building a secure TPEKS from anonymous and robust IBTD.

3.1   TPEKS Definitions and Security Model

Our model for TPEKS assumes a sender who encrypts keywords and at least t out of n servers, each equipped with its own trapdoor share, who participate in the search procedure. The latter represents the main difference to single-server based PEKS construction. We stress that the parameters t and n need not be fixed during the setup phase but can be chosen upon the generation of the trapdoors, which allows for greater flexibility.

Definition 7 (TPEKS). *A TPEKS scheme consists of the following five algorithms* (Setup, PEKS, Trpd, ShareTrpd, Test):

Setup(1^k): *On input 1^k, it generates a private/public key pair (sk, pk).*

PEKS(pk, w): *On input pk and a keyword w, it outputs a PEKS ciphertext Φ.*

ShareTrpd(pk, sk, w, t, n): *On input (pk, sk) and a keyword w, it generates a list of trapdoor shares $T_w := \{(1, T_{w,1}), \ldots, (n, T_{w,n})\}$.*

ShareTest($pk, (i, T_{w,i}), \Phi$): *On input pk, a trapdoor share $T_{w,i}$, and a PEKS ciphertext Φ, it outputs a test share τ_i.*

Test($pk, \{\tau_i\}_{i \in \Omega}, \Phi$): *On input pk, a set of test shares $\{\tau_i\}_{i \in \Omega}$, $|\Omega| \geq t$ and a PEKS ciphertext $\Phi(w')$, it outputs 1 if Φ encrypts w; otherwise it outputs 0.*

In Definition 8 we define TPEKS indistinguishability against chosen-ciphertext attacks, denoted by TPEKS-IND-CCA, aiming to protect privacy of the encrypted keywords in presence of an attacker who may learn up to $t-1$ trapdoor shares. Apart from the access to the trapdoor share oracle our scheme allows the adversary to issue up to $t-1$ queries to the test oracle.

Definition 8 (TPEKS Indistinguishability). *Let $\mathcal{B}_{ind}$ be a PPT adversary against the TPEKS-IND-CCA security of the TPEKS scheme, associated with the following experiment $\mathbf{Exp}_{\mathcal{B}_{ind}}^{TPEKS\text{-}IND\text{-}CCA\text{-}b}(1^\lambda)$:*

1. $(pk, sk) \xleftarrow{r} \mathsf{Setup}(1^\lambda, t, n)$.
2. *Let* **List** *be a list storing a keyword w and a set $T_w = \{(1, T_{w,1}), \ldots, (n, T_{w,n})\}$, where $(i, T_{w,i})$ are the outputs of $\mathsf{ShareTrpd}(pk, sk, w, t, n)$ algorithm.*
   At the beginning of the experiment the list is empty.
3. $(w_0, w_1, state) \xleftarrow{r} \mathcal{B}_{ind}^{\mathcal{O}\mathsf{ShareTrpd}(\cdot), \mathcal{O}\mathsf{Test}(\cdot)}(find, pk)$
4. *If $(w_0, T_0) \notin$ **List**, run $T_0 := \{(1, T_{w_0,1}), \ldots, (n, T_{w_0,n})\} \xleftarrow{r} \mathsf{ShareTrpd}(pk, sk, w_0, t, n)$, add (w_0, T_0) to **List***
5. *If $(w_1, T_1) \notin$ **List**, run $T_1 := \{(1, T_{w_1,1}), \ldots, (n, T_{w_1,n})\} \xleftarrow{r} \mathsf{ShareTrpd}(pk, sk, w_1, t, n)$, add (w_1, T_1) to **List***
6. $b \xleftarrow{r} \{0,1\}$; $\Phi \xleftarrow{r} \mathsf{PEKS}(pk, w_b)$
7. $b' \xleftarrow{r} \mathcal{B}_{ind}^{\mathcal{O}\mathsf{ShareTrpd}(\cdot), \mathcal{O}\mathsf{Test}(\cdot)}(guess, \Phi, state)$
   The experiment outputs 1 if all of the following holds:
   - $b' = b$
   - $\mathcal{B}_{ind}$ *asked at most $t-1$ queries to $\mathcal{O}\mathsf{ShareTrpd}(w_0, i)$ and at most $t-1$ queries to $\mathcal{O}\mathsf{ShareTrpd}(w_1, i)$*
   - $\mathcal{B}_{ind}$ *didn't query $\mathcal{O}\mathsf{Test}(w_0, \Phi)$ or $\mathcal{O}\mathsf{Test}(w_1, \Phi)$*

where the two oracles are defined as follows:

$\mathcal{O}\mathsf{ShareTrpd}(w, i)$: *On input (w, i) check whether $(w, T_w) \in$ **List**. If so, parse T_w as $\{(1, T_{w,1}), \ldots, (n, T_{w,n})\}$ and output $(i, T_{w,i})$. If $(w, T_w) \notin$ **List**, run $T_w \xleftarrow{r} \mathsf{ShareTrpd}(pk, sk, w, t, n)$. Add (w, T_w) to **List**, output $(i, T_{w,i})$.*

$\mathcal{O}\mathsf{Test}(w, \Phi)$: *On input (w, Φ) check whether $(w, T_w) \in$ **List**. If so, parse T_w as $\{(1, T_{w,i}), \ldots, (n, T_{w,n})\}$, compute $\tau_i \xleftarrow{r} \mathsf{ShareTest}(pk, (i, T_{w,i}), \Phi)$. Take at least t-out-of-n test shares τ_i, run $\mathsf{Test}(pk, \{\tau_i\}_{i \in \Omega}, \Phi)$, where $|\Omega| \geq t$, output 1 or 0. If $(w, T_w) \notin$ **List**, compute $T_w \xleftarrow{r} \mathsf{ShareTrpd}(pk, sk, w, t, n)$, add (w, T) to the **List**. Compute $\tau_i \xleftarrow{r} \mathsf{ShareTest}(pk, (i, T_{w,i}), \Phi)$, where $i \in [n]$. Take at least t-out-of-n test shares τ_i, run $\mathsf{Test}(pk, \{\tau_i\}_{i \in \Omega}, \Phi)$, output 1 or 0.*

The advantage of $\mathcal{B}_{ind}$ is defined as

$$\mathbf{Adv}_{\mathcal{B}_{ind}}^{TPEKS\text{-}IND\text{-}CCA\text{-}b}(1^k) = \left| \Pr\left[\mathbf{Exp}_{\mathcal{B}_{ind}}^{TPEKS\text{-}IND\text{-}CCA\text{-}1}(1^\lambda) = 1 \right] \right.$$
$$\left. - \Pr\left[\mathbf{Exp}_{\mathcal{B}_{ind}}^{TPEKS\text{-}IND\text{-}CCA\text{-}0}(1^\lambda) = 1 \right] \right|.$$

The scheme is TPEKS-IND-CCA secure if $\mathbf{Adv}_{\mathcal{B}_{ind}}^{TPEKS\text{-}IND\text{-}CCA}$ is negligible.

In Definition 9 we model computational consistency of TPEKS schemes, by extending the corresponding property for PEKS schemes from [1, Sect. 3]. Our definition allows the adversary to test polynomially-many keyword-ciphertext pairs, thus modeling chosen-ciphertext attacks, and accounts for the threshold setting by allowing the adversary to learn up to $t-1$ trapdoor shares for keywords that will be used to mount a successful attack.

Definition 9 (TPEKS Consistency). *Let $\mathcal{B}_c$ be a PPT adversary against the TPEKS-CONS security of the TPEKS scheme, associated with the following experiment $\mathbf{Exp}_{\mathcal{B}_c}^{\text{TPEKS-CONS}}(1^\lambda)$:*

1. $(pk, sk) \xleftarrow{r} \mathsf{Setup}(1^\lambda, t, n)$
   Let **List** *be list storing a keyword w and a set $T_w = \{(1, T_{w,1}), \ldots, (n, T_{w,n})\}$, where $(i, T_{w,i})$ are the outputs of $\mathsf{ShareTrpd}(pk, sk, w, t, n)$ algorithm.*
   At the beginning of the experiment the list is empty.
2. $(w, w') \xleftarrow{r} \mathcal{B}_c^{\mathcal{O}\mathsf{ShareTrpd}(\cdot), \mathcal{O}\mathsf{Test}(\cdot)}(pk)$
3. $\Phi \xleftarrow{r} \mathsf{PEKS}(pk, w)$
4. *If $(w', (i, T_{w',i})) \notin$* **List** *run $T_{w'} := \{(1, T_{w',1}), \ldots, (n, T_{w',n})\} \xleftarrow{r} \mathsf{ShareTrpd}(pk, sk, w', t, n)$ and add (w', T') to* **List**
   The experiment outputs 1 if all of the following holds:
   - $w \neq w'$
   - $\mathcal{B}_c$ *asked at most $t-1$ queries to $\mathcal{O}\mathsf{ShareTrpd}(w', i)$ and at most $t-1$ queries to $\mathcal{O}\mathsf{ShareTrpd}(w, i)$.*

where the two oracles are defined as follows:

$\mathcal{O}\mathsf{ShareTrpd}(w, i)$: *On input (w, i) check whether $(w, T_w) \in$* **List***. If so, parse T as $\{(1, T_{w,1}), \ldots, (n, T_{w,n})\}$ and output $(i, T_{w,i})$. If $(w, T_w) \notin$* **List***, run $T \xleftarrow{r} \mathsf{ShareTrpd}(pk, sk, w, t, n)$. Add (w, T_w) to* **List***, output $(i, T_{w,i})$.*

$\mathcal{O}\mathsf{Test}(w, \Phi)$: *On input (w, Φ) check whether $(w, T_w) \in$* **List***. If so, parse T_w as $\{(1, T_{w,i}), \ldots, (n, T_{w,n})\}$, compute $\tau_i \xleftarrow{r} \mathsf{ShareTest}(pk, (i, T_{w,i}), \Phi)$. Take at least t-out-of-n test shares τ_i, run $\mathsf{Test}(pk, \{\tau_i\}_{i \in \Omega}, \Phi)$, where $|\Omega| \geq t$, output 1 or 0. If $(w, T_w) \notin$* **List***, compute $T_w \xleftarrow{r} \mathsf{ShareTrpd}(pk, sk, w, t, n)$, add (w, T_w) to the* **List***. Compute $\tau_i \xleftarrow{r} \mathsf{ShareTest}(pk, (i, T_{w,i}), \Phi)$, where $i \in [n]$. Take at least t-out-of-n test shares τ_i, run $\mathsf{Test}(pk, \{\tau_i\}_{i \in \Omega}, \Phi)$, output 1 or 0.*

The advantage of $\mathcal{B}_c$ is defined as

$$\mathbf{Adv}_{\mathcal{B}_c}^{\text{TPEKS-CONS}}(1^k) = Pr\left[\mathbf{Exp}_{\mathcal{B}_c}^{\text{TPEKS-CONS}}(1^\lambda) = 1\right].$$

The scheme is TPEKS-CONS secure if $\mathbf{Adv}_{\mathcal{B}_c}^{\text{TPEKS-CONS}}$ is negligible.

Note: It is obvious that the correctness property of our TPEKS is satisfied. Correctness ensures that the test algorithm always outputs the correct answer.

3.2   A General TPEKS Construction from an Anonymous and Robust IBTD Scheme

The design rationale of our TPEKS construction from IBTD follows the transformation from [1] for single-server PEKS from any anonymous and robust IBE, where the PEKS private/public key pair (sk, pk) corresponds to the IBE master private/public key pair (msk, mpk) and a PEKS ciphertext $\Phi = (C, R)$ for some keyword w is computed by encrypting some random message R using keyword w as an identity. The search procedure for Φ decrypts C using the trapdoor T_w and compares the decrypted message with R.

Our IBTD-to-TPEKS transformation, detailed in Definition 10, treats TPEKS keywords as IBTD identities and performs distributed search on TPEKS ciphertexts using the IBTD threshold decryption procedure. While we describe only general IBTD-to-TPEKS transform, we remark that a concrete TPEKS instantiation can be easily obtained using our concrete IBTD construction from Definition 5.

Definition 10 (IBTD-to-TPEKS Transform).

$\mathtt{Setup}(1^k)$: *On input a security parameter 1^k, it runs the parameter generation algorithm of the IBTD scheme $(msk, mpk) \xleftarrow{r} \mathtt{Setup}(n, t, 1^k)$ and outputs $sk = msk$ and $pk = mpk$.*

$\mathtt{PEKS}(pk, w)$: *On input a public key pk and a keyword w, it runs the encryption algorithm of the IBTD scheme $C \xleftarrow{r} \mathtt{Enc}(pk, w, R)$, where $R \xleftarrow{r} \{0, 1\}^k$ is picked randomly. It returns the TPEKS ciphertext $\Phi = (C, R)$.*

$\mathtt{ShareTrpd}(pk, sk, w, t, n)$: *On input (pk, sk, w, t, n), it runs the key derivation procedure $(T_w, \{i, T_{w,i}\}) \xleftarrow{r} \mathtt{KeyDer}(pk, sk, w, t, n)$ of the IBTD scheme where sk is the master key msk and keyword w is used as id. The trapdoor T_w associated to w corresponds to sk_{id} generated by the IBTD. It outputs $\{(1, T_{w,1}), \ldots, (n, T_{w_n})\}$ which correspond to $\{(i, sk_{id,i}), \ldots, (n, sk_{id,n})\}$ of the IBTD scheme.*

$\mathtt{ShareTest}(pk, (i, T_{w,i}), \Phi)$: *On input pk, a trapdoor share $(i, T_{w,i})$, and a TPEKS ciphertext $\Phi = (C, R)$, it outputs $\tau_i \xleftarrow{r} \mathtt{ShareDec}(pk, (i, T_{w,i}), \Phi)$ using the distributed decryption algorithm of the IBTD scheme.*

$\mathtt{Test}(pk, \{\tau_{w,i}\}_{i \in \Omega}, (C, R))$: *On input of at least t test shares $\{\tau_{w,i}\}_{i \in \Omega}$, $|\Omega| \geq t$ and (C, R) is a TPEKS ciphertext, it computes $R' \xleftarrow{r} \mathtt{Dec}(pk, \{\delta_i\}_{i \in \Omega}, C)$ and outputs 1 if $R' = R$, and 0 otherwise.*

3.3   Security Analysis

With regard to security, in Theorem 4, we establish similar implications for TPEKS indistinguishability as in [1], namely we rely on the anonymity and robustness of the IBTD scheme. In Theorem 5 we show that for TPEKS consistency the underlying IBTD scheme must not only be indistinguishable but also robust. This contrasts to [1,2] where IBE robustness was not required for

PEKS consistency and is mainly due to the fact that our definition of consistency allows adaptive queries to the distributed test procedure, which was not an issue in [1,2].

Theorem 4 (TPEKS-IND-CCA). *If IBTD scheme is* IBTD-ANO-CCA *secure then the obtained TPEKS scheme in* Definition 10 *is* TPEKS-IND-CCA *secure*

Proof. We use a TPEKS-IND-CCA adversary $\mathcal{B}_{ind}$ against the TPEKS scheme to construct a simulator $\mathcal{S}$ that breaks the assumed IBTD-ANO-CCA and IBTD-SROB-CCA properties of the IBTD scheme. That is, $\mathcal{S}$ acts as $\mathcal{A}_{ano}$ attacking the IBTD-ANO-CCA security and as $\mathcal{A}_{rob}$ against IBTD-SROB-CCA. The simulation of the view of $\mathcal{B}_{ind}$ happens in several games. The initial game is the Game0 which describes the real attack. First the challenger runs the **Setup** of the TPEKS scheme on input a security parameter λ, threshold parameter t and number of servers n. The challenger gives $\mathcal{B}_{ind}$ the public key pk. If $\mathcal{B}_{ind}$ submits a pair of keywords w_0, w_1, the challenger computes a target ciphertext Φ^*. $\mathcal{B}_{ind}$ issues trapdoor share and test queries on the PEKS ciphertext Φ. The first game (Game1) differs from the previous one by simulation of trapdoor share queries. If these shares are involved in computing the PEKS ciphertext, the simulator modifies the challenge ciphertext. The rest of Game 0 remains unmodified. The simulation is distributed in two subcases. In Case 1 holds $C \neq C^*, w^* \neq w_b$, which invokes $\mathcal{A}_{ano}$ to simulate the queries on identities $id \neq id_b$. In Case 2 holds $C = C^*$, $w^* = w_b$ for $b \in \{0,1\}$ where $\mathcal{A}_{rob}$ is invoked for the simulation of queries on id^*. In Case 1, $\mathcal{A}_{ano}$ aborts the game if $\mathcal{B}_{ind}$ issues more than $t-1$ queries on $w^* \neq w_b$. In Case 2, where holds $C = C^*$, $\mathcal{A}_{rob}$ aborts the game if $\mathcal{B}_{ind}$ issues more than $t-1$ queries on $w^* = w_b = w_{1-b}$. The second game differs from Game 1 by simulation of test queries. In Case 1, $\mathcal{A}_{ano}$ aborts the game if $\mathcal{B}_{ind}$ issues queries on w_b. In Case 2, where holds $C = C^*$, $\mathcal{A}_{rob}$ aborts the game if $\mathcal{B}_{ind}$ issues queries on $w^* = w_b = w_{1-b}$. Each of the simulation steps looks as follows:

Setup: $\mathcal{S}$ is given as input mpk of IBTD. It sets TPEKS public key pk equal to mpk. We assume that a set of $t-1$ servers have been corrupted. When $\mathcal{B}_{ind}$ issues trapdoor share and decryption queries on input (w, i) and (w, Φ) respectively, where $w \neq w_b$, and $\Phi = (C, R), \Phi^* = (C^*, R^*)$, $\mathcal{S}$ distinguishes between two cases - Case 1: $C \neq C^*, w \neq w_b$ where $\mathcal{A}_{ano}$ is invoked and Case 2: $C = C^*$, $w = w_b$ for $b \in \{0,1\}$ where $\mathcal{A}_{rob}$ is invoked.

Queries to $\mathcal{O}\mathsf{ShareTrpd}$: Let (w, i) be a trapdoor share query issued by $\mathcal{B}_{ind}$. $\mathcal{S}$ sets $w \leftarrow id$ and queries its oracle $\mathcal{O}\mathsf{KeyDer}(w, i)$. The oracle outputs $(i, sk_{w,i})$, which $\mathcal{S}$ sets equal to $(i, T_{w,i})$ and returns it to $\mathcal{B}_{ind}$. In Case 1, the simulator represented by $\mathcal{A}_{ano}$ aborts if $\mathcal{B}_{ind}$ issued more than $t-1$ queries to $\mathcal{O}\mathsf{ShareTrpd}(w_0, i)$ and to $\mathcal{O}\mathsf{ShareTrpd}(w_1, i)$. In Case 2, simulator represented by $\mathcal{A}_{rob}$ aborts if $w_0 = w_1$ or $(w_0, T_0, I_0) \notin \mathbf{List}$, or $(w_1, T_1, I_1) \notin \mathbf{List}$, or $|I_0| \geq t, |I_1| \geq t$, where $T_b = \{(1, T_{w_b,1}), \ldots, (n, T_{w_b,n})\}, I_b \subset [1, n], b \in \{0, 1\}$.

Queries to $\mathcal{O}\mathsf{Test}$: $\mathcal{B}_{ind}$ issues test queries on (w, Φ). $\mathcal{S}$ sets $w \leftarrow id, \Phi = (C, R)$ and queries its oracle $\mathcal{O}\mathsf{Dec}(w, C)$. The oracle outputs m. $\mathcal{S}$ sets $R \leftarrow m$ and returns R to $\mathcal{B}_{ind}$. In Case 1, $\mathcal{S}$ aborts simulation if $\mathcal{B}_{ind}$ queried $\mathcal{O}\mathsf{Test}(w_0, \Phi^*)$ or $\mathcal{O}\mathsf{Test}(w_1, \Phi^*)$, where $\Phi^* = (C^*, R^*)$. In Case 2, $\mathcal{S}$ aborts if $w_0 = w_1$ or $(w_0, T_0, I_0) \notin \mathbf{List}$, or $(w_1, T_1, I_1) \notin \mathbf{List}$, or $|I_0| \geq t, |I_1| \geq t$, where $T_b = \{(1, T_{w_b,1}), \ldots, (n, T_{w_b,n})\}$ and $I_b \subset [1, n]$, for $b \in \{0, 1\}$.

Challenge: $\mathcal{B}_{ind}$ outputs two identities w_0, w_1 and $R^* \xleftarrow{r} \{0, 1\}^{\ell}$. $\mathcal{S}$ responds with IBTD ciphertext $\Phi^* = (C^*, R^*)$, where $C^* \leftarrow \mathsf{Enc}(pk, w_b, R^*)$, for $b \in \{0, 1\}$ and $R^* \xleftarrow{r} \{0, 1\}$. $\mathcal{S}$ returns its ciphertext $C^* \leftarrow Enc(mpk, w_b, R^*)$.

Analysis: In Case 1: Simulator $\mathcal{S}$ is represented by $\mathcal{A}_{ano}$. Let q_{ts}, q_t be the number of issued trapdoor share queries on different id's. We assume that $\mathcal{B}_{ind}$ corrupts $t - 1$-out-of-n servers with index $i_1, \ldots, i_{t-1}$. The probability that $\mathcal{S}$ corrupts a server j with $j \in \{i_1, \ldots, i_{t-1}\}$ is $1/\binom{n}{t-1}$. Let E denote the event that $\mathcal{B}_{ind}$ wins the indistinguishability experiment from Definition 8. Let $E1$ denote the event that $\mathcal{B}_{ind}$ wins Game 1. It holds $\frac{1}{2}\mathbf{Adv}_{\mathcal{B}_{ind}}^{\text{TPEKS-IND-CCA}}(1^\lambda) = Pr[E] - 1/2 \geq 1/\binom{n}{t-1}\frac{1}{q_{ts}}(Pr[E1] - 1/2)$. Let $E2$ denote the event that $\mathcal{B}_{ind}$ wins Game 2, then holds: $Pr[E1] - 1/2 \geq \frac{1}{q_t}(Pr[E2] - 1/2)$

$\Leftrightarrow \frac{1}{2}\mathbf{Adv}_{\mathcal{B}_{ind}}^{\text{TPEKS-IND-CCA}}(1^\lambda) = Pr[E] - 1/2 \geq 1/\binom{n}{t-1}\frac{1}{q_{ts}}\frac{1}{q_t}(Pr[E2] - 1/2)$ In Case 2: Simulator $\mathcal{S}$ is represented by $\mathcal{A}_{rob}$. Let q'_{ts}, q'_t be the number of issued trapdoor share queries on different id's. We assume that $\mathcal{B}_{ind}$ corrupts $t - 1$-out-of-n servers with index $i_1, \ldots, i_{t-1}$. The probability that $\mathcal{S}$ corrupts a server j with $j \in \{i_1, \ldots, i_{t-1}\}$ is $1/\binom{n}{t-1}$. Let E denote the event that $\mathcal{B}_{ind}$ wins the indistinguishability experiment from Definition 8. Let $\widetilde{E1}$ denote the event that $\mathcal{B}_{ind}$ wins Game 1. It holds $\frac{1}{2}\mathbf{Adv}_{\mathcal{B}_{ind}}^{\text{TPEKS-IND-CCA}}(1^\lambda) = Pr[E] - 1/2 \geq 1/\binom{n}{t-1}\frac{1}{q_{ts}}(Pr[\widetilde{E1}] - 1/2)$. Let $\widetilde{E2}$ denote the event that $\mathcal{B}_{ind}$ wins Game 2, then holds: $Pr[E1] - 1/2 \geq \frac{1}{q_t}(Pr[\widetilde{E2}] - 1/2)$

$\Leftrightarrow \frac{1}{2}\mathbf{Adv}_{\mathcal{B}_{ind}}^{\text{TPEKS-IND-CCA}}(1^\lambda) = Pr[E] - 1/2 \geq 1/\binom{n}{t-1}\frac{1}{q_{ts}}\frac{1}{q_t}(Pr[\widetilde{E2}] - 1/2)$ The total advantage of $\mathcal{B}_{ind}$ is given by

$$\frac{1}{2}\mathbf{Adv}_{\mathcal{B}_{ind}}^{\text{TPEKS-IND-CCA}}(1^\lambda) = Pr[E|Case1] + Pr[E|Case2] = 2Pr[E] - 1$$

$$\geq 1/\binom{n}{t-1}\frac{1}{q_{ts}}\frac{1}{q_t}(Pr[E2] - 1/2) + 1/\binom{n}{t-1}\frac{1}{q_{ts}}\frac{1}{q_t}(Pr[\widetilde{E2}] - 1/2)$$

$$= 1/\binom{n}{t-1}\frac{1}{q_{ts}}\frac{1}{q_t}(Pr[E2] + Pr[\widetilde{E2}] - 1)$$

Theorem 5 (TPEKS Consistency). *If IBTD scheme is IBTD-IND-CCA secure then the obtained TPEKS scheme in Definition 10 is TPEKS-CONS secure.*

Proof. For the proof of this theorem we refer to Appendix B.

3.4   Application to Cloud Setting

In the introduction we mentioned the applicability of TPEKS to distributed cloud storage, as a solution to mitigate the single point of trust with regard to the search procedure and the insecurity of single-server PEKS schemes against keyword guessing attacks. Together with its properties, TPEKS seems to be particularly attractive for this application, as detailed in the following two scenarios.

In the first use case we assume an user who wishes to upload his data files on the cloud servers to have access to these file at a later point in time. Assume an user who uploads to n cloud servers m encrypted data files with m PEKS ciphertexts where each of them is encrypted on l different keywords $w_1, \ldots, w_l$, i.e. PEKS ciphertext for the j-th file is given by $\{\Phi(pk, j, w_{i_1}, \ldots, w_{i_d})\}_{i_d \in [l], j \in [m]}$. When the user wants to download files which contain a keyword w_i, he computes trapdoor shares for each server on that keyword, i.e. he sends trapdoors $T_{w_i, 1}, \ldots, T_{w_i, n}$ on w_i to the n servers, where $i \in \{1, \ldots, l\}$ denotes the index one of the l keywords. Each cloud server computes test shares taking as input the different PEKS ciphertext for each data file and the trapdoor for the k-th server $\tau_{k,j} \leftarrow \mathsf{ShareTest}(\Phi(pk, j, w_{i_1}, \ldots, w_{i_d}), T_{w_i, k})$, where j denotes the index of PEKS ciphtertext for file j and $i_1, \ldots, i_d \in [l]$ is a set of l keywords. Each server outputs m test shares $\{\tau_{k,j}\}_{j \in [m]}$, such that the user obtains in total $m \times n$ different test shares. For the ease of analysis we observe 2 cloud servers and therefore $2m$ different test shares. Since the user does not know which test shares belong to which files, he has to run m^2 test algorithms, namely $\mathsf{Test}(\tau_{1,j}, \tau_{2,j'}, pk)$ where $j, j' \in [m]$. This scenario has a total complexity of $\mathcal{O}(m^2)$. This scenario guarantees privacy of the user, because the trapdoors do not reveal anything about the keywords and the servers do not learn anything about the keywords since they take the ciphertexts keywords and the trapdoors without getting any information about the content of the inputs.

To reduce the complexity the user could use random indices for each uploaded PEKS ciphertexts on keywords. That means that he would need to upload $(r_\chi, \{\Phi(pk, j, w_{i_1}, \ldots, w_{i_d})\}_{i_d \in [l], j \in [m]})$ to each server, where r_χ denotes a random index for the ciphertext $\Phi(\cdot)$ on a set of keywords $w_{i_1}, \ldots, w_{i_d}$, with $i_1, \ldots, i_d \in [l]$. The user has to remember $(r_\chi, w_{i_1}, \ldots, w_{i_d})$ for later use. Finally if he wants to download data files with keyword $w_i, i \in [l]$ he computes n trapdoor shares for all n servers and sends them together with the index r_χ to the server. Each server compares whether the received randomness belongs to one of the stored PEKS ciphertexts. If so each server computes trapdoor shares $(r_\chi, \tau_{k,j}) \leftarrow \mathsf{ShareTest}((r_\chi, \Phi(pk, j, w_{i_1}, \ldots, w_{i_d}), T_{w_i, k})$ and sends them to the user. Upon receiving the test shares together with the randomness, the user can recognize which test shares to which file and can be used to run test algorithm. If the output of the algorithm is 1, the user sends the randomness to one of the servers to get access for the download of a file. The complexity in this scenario can be reduced to the linear size $\mathcal{O}(m)$. This example still guarantees privacy of the user because a randomness is prepared for a set of keywords, such that the files are unlinkable to the keywords.

As a second use case we consider a sender who sends a set of m encrypted messages with PEKS ciphertexts $\Phi(pk, j, w_1, \ldots, w_l)$ on l different keywords to the n servers, where $j \in [m]$, for a recipient who owns the public key pk. The receipient computes trapdoor shares for each server on a set of required keywords and sends them to the servers. Each of the servers computes m different test shares and return them to the user. The user needs to find the sets of the test shares for the same encrypted messages. To do so he needs to try m^n combinations which gives us the complexity of this scenario, $\mathcal{O}(m^n)$. To reduce the complexity the user could compute $l \times n$ trapdoor shares for the l shares and send them together with a randomness $\{r_\chi\}_{\chi \in [l]}$ of each keyword to the servers. Each of the servers runs m test share algorithms and returns $m \times l$ test shares together with randomness such that the user can combine the test shares with the corresponding randomness. The complexity of this scenario is reduced to $\mathcal{O}(l \times m)$. The privacy of user regarding the keywords remains guaranteed, because no one can learn the queried keywords.

A   Proof of Theorem 1

Proof. Let $\mathcal{A}_{ind}$ be an adversary that defeats the IBTD-IND-CCA security of the IBTD scheme, and $\mathcal{A}_{BDH}$ let be an adversary for the BDH problem. $\mathcal{A}_{BDH}$ are given BDH parameters $(\mathbb{G}, \mathbb{G}_T, e, q)$ and a random instance $(g, g^a, g^b, g^c, e(g, g)^{abc})$ of the BDH problem for these parameters. That means $g \xleftarrow{r} \mathbb{G}^*$ and $a, b, c \xleftarrow{r} \mathbb{Z}_q^*$ are random. Let $Z = e(g, g)^z \in \mathbb{G}_T$, where $\mathcal{A}_{ind}$'s aim is to decide whether $z = abc$, or z is a randomly chosen value from $\mathbb{Z}_q^*$. The definition of CCA security allows the adversary to obtain the secret share associated with any identity id_i, where $i \in [q_1, \ldots, q_m]$ of her choice and $\mathcal{A}_{ind}$ is challenged on a public key id^* of her choice. $\mathcal{A}_{ind}$ issues queries $q_1, \ldots, q_m$, where q_i, for $i \in [m]$ is one the key derivation or decryption queries. $\mathcal{A}_{BDH}$ uses $\mathcal{A}_{ind}$ to find d as follows:

Setup: The simulator $\mathcal{A}_{BDH}$ generates IBTD master public key $mpk = (\mathbb{G}, \mathbb{G}_T, q, g, e, H_1, H_2, H_3, H_4, Y)$ by setting $Y = g^c$ and $Q_{id} = g^b$. $H_1, \ldots, H_4$ are random oracles controlled by $\mathcal{A}_{BDH}$. The queries to the oracles are described as follows. $\mathcal{A}_{BDH}$ gives mpk to $\mathcal{A}_{ind}$. $\mathcal{A}_{ind}$ issues q_{H_1}, q_{ks} queries on an id to H_1 oracle, the key derivation oracle and a query on id^* to the both oracles in the challenge stage, respectively.

H_1 **Oracle Queries:** Let H_1 List be a list used for storing the results of queries to the H_1 oracle, (id, Q_{id}). Whenever H_1 is queried at $id \in \{0, 1\}^\ell$, $\mathcal{A}_{BDH}$ does the following: If $(id, Q_{id}) \in H_1$List, it returns Q_{id} to $\mathcal{A}_{ind}$. For $id \neq id^*$, $\mathcal{A}_{BDH}$ sets $Q_{id} = H_1(id) = (g^b)^\gamma$ for a random $\gamma \xleftarrow{r} \mathbb{Z}_q$, where g^b is given from the BDH instance. If $id = id^*$, $\mathcal{A}_{BDH}$ computes $Q_{id} = g^\gamma$, where $\gamma \xleftarrow{r} \mathbb{Z}_q$, adds it to the H_1List and returns Q_{id} to $\mathcal{A}_{ind}$.

H_2 **Oracle Queries:** Let H_2 List be a list consisting of all pairs $(\kappa_{id}, H_2(\kappa_{id}))$. When $\mathcal{A}_{ind}$ queries on input $\kappa_{id} = e(Q_{id}, Y)^r$, $\mathcal{A}_{BDH}$ checks whether the queried value is in the H_2List, if so it returns $H_2(\kappa_{id})$. Otherwise it chooses a random $H_2' \in \{0, 1\}^\ell$ and gives $H_2(\kappa_{id}) = H_2'$ to $\mathcal{A}_{ind}$.

H_3 **Oracle Queries:** Let H_3 List consisting of elements $(\sigma, m, H_3(\sigma, m))$, where $\sigma \in_r \{0, 1\}^l$. When $\mathcal{A}_{ind}$ issues queries on input (σ, m) it invokes $\mathcal{A}_{BDH}$ that checks whether $(\sigma, m) \in H_3$List. If so it returns the corresponding $H_3(\sigma, m)$.

Otherwise $\mathcal{A}_{BDH}$ chooses a random $H_3' \in \mathbb{Z}_q^*$ and sets $H_3' = H_3(\sigma, m)$ that it gives to $\mathcal{A}_{ind}$

H_4 **Oracle Queries:** Let H_4 List consist of all pairs $(\sigma, H_4(\sigma))$. When $\mathcal{A}_{ind}$ issues a query on $(\sigma, \cdot)$, $\mathcal{A}_{BDH}$ checks, whether $\sigma \in H_4$List. If so, it returns the corresponding $H_4(\sigma)$, otherwise it chooses $H_4' \in \{0,1\}^\ell$ and sets $H_4' = H_4(\sigma)$ and gives H_4' to $\mathcal{A}_{ind}$.

Phase 1: $\mathcal{A}_{ind}$ issues up to q_m queries to the key derivation and decryption oracles.

Queries to $\mathcal{O}$KeyDer(id, i): For $id \neq id^*$: When $\mathcal{A}$ submits a key derivation query on input (id, i), $\mathcal{A}_{BDH}$ checks whether $(id, S) \in$ **List**. If so, $\mathcal{A}_{BDH}$ returns the corresponding secret share $sk_{id,i}$ for index i to $\mathcal{A}_{ind}$. If $(id, S) \notin$ **List**, $\mathcal{A}_{BDH}$ simulates the key shares as follows: For the $t-1$ corrupted servers with indices $i_1, \ldots, i_{t-1}$, $\mathcal{A}_{BDH}$ chooses $t-1$ random values χ_i, such that $Q_{id}^{f(i)} = \chi_i$. If $id = id^*$, $\mathcal{A}_{BDH}$ picks a random $\chi_i \in \mathbb{G}$ and returns it to $\mathcal{A}_{ind}$. If $\mathcal{A}_{ind}$ issues more than $t-1$ queries on id^*, $\mathcal{A}_{BDH}$ aborts the simulation. If $id \neq id^*$, $\mathcal{A}_{BDH}$ chooses $\gamma \xleftarrow{r} \mathbb{Z}_q^*$, adds $Q_{id} = (g^b)^\gamma$ to the H_1List and outputs to $\mathcal{A}_{ind}$

Queries to $\mathcal{O}$Dec(id, C): If $id = id^*$, $\mathcal{A}_{BDH}$ aborts the simulaton. For $id \neq id^*$, $\mathcal{A}_{ind}$ issues a decryption query on input (id, C) to its decryption oracle, where $C \neq C^*$. $\mathcal{A}_{BDH}$ simulates the decryption oracle without knowing the decryption shares. It does the following:

1. $\mathcal{A}_{BDH}$ checks whether $(id, H_1(id)) \in H_1$list. If so, it fixes the corresponding $Q_{id} = H_1(id)$.
2. It computes $\kappa_{id} = e(Q_{id}, Y)^r$, using the fixed Q_{id} from H_1List and $Y = g^c$.
3. To determine the corresponding r, the simulator searches the H_3List. Choosing each triple (σ, m, r), the simulator compares, whether $g^r = U$, where U is given from the received ciphertext. After fixing the matching r, $\mathcal{A}_{BDH}$ receives (σ, m).
4. Using the fixed r and σ from the previous step the simulator computes $\kappa_{id} = e(Q_{id}, Y)^r$ and searches the H_2List for the corresponding value $H_2(k_{id})$. It checks, whether $V = \sigma \oplus H_2(\kappa_{id})$.
5. Taking σ, m from step 3. the simulator searches H_4List for the corresponding $H_4(\sigma)$ entry. Upon finding the matching value it checks whether $W = m \oplus H_4(\sigma)$. If one of the computations in the above 5 steps fails, the simulator aborts the game. Otherwise if all 5 steps finished successful, $\mathcal{A}_{BDH}$ returns m to $\mathcal{A}_{ind}$.

Challenge Ciphertext: At some point $\mathcal{A}_{ind}$ outputs two messages m_0, m_1 and an identity id on which it wishes to be challenged. $\mathcal{A}_{BDH}$ simulates the ciphertext as follows. He replaces U by g^a from its BDH instance. It sets $Q_{id} = g^b$ and $Y = g^c$ such that $e(Q_{id}, Y) = e(g^b, g^c)$ and $\kappa_{id} = Z$, where Z is the value from BDH instance. It chooses $s \in \{0,1\}^\ell$ uniformly at random. $\mathcal{A}_{BDH}$ gives $C^* = (g^a, s \oplus H_2(Z), m_\beta \oplus H_4(s)), \beta \in \{0,1\}$ as challenge to $\mathcal{A}$.

Phase 2: $\mathcal{A}_{ind}$ issues additional queries as in Phase 1, to which $\mathcal{A}_{BDH}$ responds as before

Guess: Eventually $\mathcal{A}_{ind}$ outputs β' as its guess for β. Algorithm $\mathcal{A}_{BDH}$ outputs β' as its guess for β

Analysis: Let q_{H_1}, q_{ks}, q_d be the number of issued H_1 oracle queries, key share queries, decryption queries on an identity id, respectively and $\mathcal{A}_{ind}$ issues one query on challenge id to the three oracles. The probability that $\mathcal{A}_{BDH}$ guesses the correct challenge id is $\delta_1 := \frac{1}{q_{H_1}+q_{ks}+1}$. It aborts the simulation with probability δ if $id = id^*$, which has already been queried either to H_1 oracle or to $\mathcal{O}$Dec. We assume that $\mathcal{A}_{ind}$ corrupts $t-1$-out-of-n servers with index $i_1, \ldots, i_{t-1}$. The probability that $\mathcal{A}_{BDH}$ matches a server j with $j \in \{i_1, \ldots, i_{t-1}\}$ is $\delta_2 := \frac{1}{\binom{n}{t-1}}$. If $\mathcal{A}_{ind}$ issues more than $t-1$ secret share queries on the same identity id, $\mathcal{A}_{BDH}$ aborts the simulation. The simulator aborts the decryption simulation in the non challenge phase with negligible probability δ_3, where $\delta_3 \in [0,1]$. The probability that it does not abort in the first phase is $1-(\delta_1+\delta_3)$. The simulator aborts in the challenge phase if $\mathcal{A}_{ind}$ issues more than $t-1$ queries to $\mathcal{O}$Dec and $\mathcal{O}$KeyDer on the challenge identity id^*. It also stops the simulation if $Z = e(g,g)^{abc}$, where δ_4 is the probability, that the equation holds. The probability for abortion during the key derivation or decryption queries on id^* is $\delta_2 + \delta_2\delta_3$. The probability that it does not abort in the challenge step is $1-(\delta_2+\delta_2\delta_3+\delta_4)$. Therefore the probability that $\mathcal{A}_{BDH}$ does not abort during the simulation is $(1-(\delta_1+\delta_3))(1-(\delta_2+\delta_2\delta_3+\delta_4)) = 1-\tilde{\delta}$, where $\tilde{\delta}$ is negligible. Advantage of $\mathcal{A}_{ind}$ is given by

$$\mathbf{Adv}_{\mathbf{A}_{\mathbf{BDH}}} \geq \mathbf{Adv}_{\mathcal{A}_{ind}}^{IBTD\text{-}IND\text{-}CCA} = \mathbf{1} - \tilde{\delta}$$

It follows that $\mathbf{Adv}_{\mathbf{A}_{\mathbf{BDH}}} > 1 - \tilde{\delta}$ is non-negligible which is a contradiction to the assumption. Therefore we follow, that the advantage of $\mathcal{A}_{ind}$ is negligible.

B   Proof of Theorem 5

Proof. We use a TPEKS-CONS adversary $\mathcal{B}_c$ to construct a simulator $\mathcal{S}$ that breaks the IBTD-IND-CCA and IBTD-SROB-CCA properties of IBTD. That is, $\mathcal{S}$ acts as $\mathcal{A}_{ind}$ against IBTD-IND-CCA and as $\mathcal{A}_{rob}$ against IBTD-SROB-CCA. The initial game is the Game 0 which describes the real attack. First the challenger runs the **Setup** of the TPEKS scheme on input a security parameter λ, threshold parameter t and number of servers n. The challenger gives $\mathcal{B}_c$ the public key pk. If $\mathcal{B}_c$ submits a pair of keywords w, w', the challenger computes a target ciphertext Φ^*. $\mathcal{B}_c$ issues trapdoor share and test queries on the PEKS ciphertext Φ. The first game (Game 1) differs from the previous one by simulation of trapdoor share queries. If these shares are involved in computing the PEKS ciphertext, the simulator modifies the challenge ciphertext. The rest of Game 0 remains unmodified. The simulation is distributed in two subcases. In Case 1 holds $C \neq C^*, w^* \neq w'$, which invokes $\mathcal{A}_{ind}$ to simulate the queries on identities $id^* \neq id'$. In Case 2 holds $C = C^*, w^* = w'$, where $\mathcal{A}_{rob}$ is invoked for the simulation of queries on id^*. In Case 1, $\mathcal{A}_{ind}$ aborts the game if $\mathcal{B}_c$ issues more than $t-1$ queries on w^* or $w^* = w'$. In Case 2, where holds $C = C^*$, $\mathcal{A}_{rob}$

aborts the game if $\mathcal{B}_c$ issues more than $t-1$ queries on $w^* = w'$. The second game differs from Game 1 by simulation of test queries. In Case 1, $\mathcal{A}_{ind}$ aborts the game if $\mathcal{B}_c$ issues queries on w^*. In Case 2, where holds $C = C^*$, $\mathcal{A}_{rob}$ aborts the game if $\mathcal{B}_c$ issues queries on $w^* = w'$. Each of the simulation steps looks as follows:

Setup: $\mathcal{S}$ is given as input mpk of IBTD. It sets TPEKS public key pk equal to mpk. We assume that a set of $t-1$ servers have been corrupted. When $\mathcal{B}_c$ issues trapdoor share and decryption queries on input (w, i) and (w, Φ) respectively, where $w \neq w'$, and $\Phi = (C, R), \Phi^* = (C^*, R^*)$, $\mathcal{S}$ distinguishes between two cases - Case 1: $C \neq C^*, w \neq w'$ where $\mathcal{A}_{ind}$ is invoked and Case 2: $C = C^*$, $w = w'$, where $\mathcal{A}_{rob}$ is invoked.

Queries to $\mathcal{O}\mathsf{ShareTrpd}$: Let (w, i) be a trapdoor share query issued by $\mathcal{B}_c$. $\mathcal{S}$ sets $w \leftarrow id$ and queries its oracle $\mathcal{O}\mathsf{KeyDer}(w, i)$. The oracle outputs $(i, sk_{w,i})$, which $\mathcal{S}$ sets equal to $(i, T_{w,i})$ and returns it to $\mathcal{B}_c$. In Case 1, the simulator represented by $\mathcal{A}_{ind}$ aborts if $\mathcal{B}_c$ issued more than $t-1$ queries to $\mathcal{O}\mathsf{ShareTrpd}(w_0, i)$ and to $\mathcal{O}\mathsf{ShareTrpd}(w_1, i)$. In Case 2, simulator represented by $\mathcal{A}_{rob}$ aborts if $w = w'$ or $(w, T, I) \notin \mathbf{List}$, or $(w', T', I') \notin \mathbf{List}$, or $|I| \geq t, |I'| \geq t$, where $T = \{(1, T_{w',1}), \ldots, (n, T_{w',n})\}, I, I' \subset [1, n]$.

Queries to $\mathcal{O}\mathsf{Test}$: $\mathcal{B}_{ind}$ issues test queries on (w, Φ). $\mathcal{S}$ sets $w \leftarrow id, \Phi = (C, R)$ and queries its oracle $\mathcal{O}\mathsf{Dec}(w, C)$. The oracle outputs m. $\mathcal{S}$ sets $R \leftarrow m$ and returns R to $\mathcal{B}_c$. In Case 1, $\mathcal{S}$ aborts simulation if $\mathcal{B}_c$ queried $\mathcal{O}\mathsf{Test}(w, \Phi^*)$ or $\mathcal{O}\mathsf{Test}(w', \Phi^*)$, where $\Phi^* = (C^*, R^*)$. In Case 2, $\mathcal{S}$ aborts if $w = w'$ or $(w', T', I') \notin \mathbf{List}$, or $(w, T, I) \notin \mathbf{List}$, or $|I| \geq t, |I'| \geq t$, where $T = \{(1, T_{w,1}), \ldots, (n, T_{w,n})\}$ and $I, I' \subset [1, n]$.

Challenge: $\mathcal{B}_c$ outputs a challenge identity w^* and $R_0, R_1 \xleftarrow{r} \{0,1\}^\ell$. $\mathcal{S}$ responds with IBTD ciphertext $\Phi^* = (C^*, R_b)$, where $C^* \leftarrow \mathsf{Enc}(pk, w^*, R_b)$, for $b \in \{0,1\}$ and $R^* \xleftarrow{r} \{0,1\}$. $\mathcal{S}$ returns its ciphertext $C^* \leftarrow Enc(mpk, w^*, Rb)$.

Analysis: In Case 1: Simulator $\mathcal{S}$ is represented by $\mathcal{A}_{inc}$. Let q_{ts}, q_t be the number of issued trapdoor share queries on different id's. We assume that $\mathcal{B}_c$ corrupts $t-1$-out-of-n servers with index $i_1, \ldots, i_{t-1}$. The probability that $\mathcal{S}$ corrupts a server j with $j \in \{i_1, \ldots, i_{t-1}\}$ is $1/\binom{n}{t-1}$. Let E denote the event that $\mathcal{B}_c$ wins the indistinguishability experiment from Definition 8. Let $E1$ denote the event that $\mathcal{B}_c$ wins Game 1. It holds $\frac{1}{2}\mathbf{Adv}_{\mathcal{B}_c}^{\text{TPEKS-CONS}}(1^\lambda) = Pr[E] - 1/2 \geq 1/\binom{n}{t-1}\frac{1}{q_{ts}}(Pr[E1] - 1/2)$. Let $E2$ denote the event that $\mathcal{B}_c$ wins Game 2, then holds:

$$Pr[E1] - 1/2 \geq \frac{1}{q_t}(Pr[E2] - 1/2)$$

$$\Leftrightarrow \frac{1}{2}\mathbf{Adv}_{\mathcal{B}_c}^{\text{TPEKS-CONS}}(1^\lambda) = Pr[E] - 1/2 \geq 1/\binom{n}{t-1}\frac{1}{q_{ts}}\frac{1}{q_t}(Pr[E2] - 1/2).$$

In Case 2: Simulator $\mathcal{S}$ is represented by $\mathcal{A}_{rob}$. Let q'_{ts}, q'_t be the number of issued trapdoor share queries on different id's. We assume that $\mathcal{B}_c$ corrupts $t-1$-out-of-n servers with index $i_1, \ldots, i_{t-1}$. The probability that $\mathcal{S}$ corrupts a server j with $j \in \{i_1, \ldots, i_{t-1}\}$ is $1/\binom{n}{t-1}$. Let E denote the event that $\mathcal{B}_c$

wins the indistinguishability experiment from Definition 8. Let $\widetilde{E1}$ denote the event that $\mathcal{B}_c$ wins Game 1. It holds $\frac{1}{2}\mathbf{Adv}_{\mathcal{B}_c}^{\text{TPEKS-CONS}}(1^\lambda) = Pr[E] - 1/2 \geq 1/\binom{n}{t-1}\frac{1}{q_{ts}}(Pr[\widetilde{E1}] - 1/2)$. Let $\widetilde{E2}$ denote the event that $\mathcal{B}_c$ wins Game 2, then holds:

$$Pr[E1] - 1/2 \geq \frac{1}{q_t}(Pr[\widetilde{E2}] - 1/2)$$

$$\Leftrightarrow \frac{1}{2}\mathbf{Adv}_{\mathcal{B}_c}^{\text{TPEKS-CONS}}(1^\lambda) = Pr[E] - 1/2 \geq 1/\binom{n}{t-1}\frac{1}{q_{ts}}\frac{1}{q_t}(Pr[\widetilde{E2}] - 1/2)$$

The total advantage of $\mathcal{B}_c$ is given by

$$\frac{1}{2}\mathbf{Adv}_{\mathcal{B}_c}^{\text{TPEKS-CONS}}(1^\lambda) = Pr[E|Case\,1] + Pr[E|Case\,2] = 2Pr[E] - 1$$

$$\geq 1/\binom{n}{t-1}\frac{1}{q_{ts}}\frac{1}{q_t}(Pr[E2] - 1/2) + 1/\binom{n}{t-1}\frac{1}{q_{ts}}\frac{1}{q_t}(Pr[\widetilde{E2}] - 1/2)$$

$$= 1/\binom{n}{t-1}\frac{1}{q_{ts}}\frac{1}{q_t}(Pr[E2] + Pr[\widetilde{E2}] - 1).$$

References

1. Abdalla, M., Bellare, M., Catalano, D., Kiltz, E., Kohno, T., Lange, T., Malone-Lee, J., Neven, G., Pailier, P., Shi, H.: Searchable encryption revisited: consistency properties, relation to anonymous IBE and extensions. J. Cryptol. **21**, 350–391 (2008)
2. Abdalla, M., Bellare, M., Neven, G.: Robust encryption. In: Micciancio, D. (ed.) TCC 2010. LNCS, vol. 5978, pp. 480–497. Springer, Heidelberg (2010)
3. Baek, J., Safavi-Naini, R., Susilo, W.: Public key encryption with keyword search revisited. IACR Cryptology ePrint Archive, p. 191 (2005)
4. Baek, J., Safavi-Naini, R., Susilo, W.: Public key encryption with keyword search revisited. In: Gervasi, O., Murgante, B., Laganà, A., Taniar, D., Mun, Y., Gavrilova, M.L. (eds.) ICCSA 2008, Part I. LNCS, vol. 5072, pp. 1249–1259. Springer, Heidelberg (2008)
5. Baek, J., Zheng, Y.: Identity-based threshold decryption. In: Bao, F., Deng, R., Zhou, J. (eds.) PKC 2004. LNCS, vol. 2947, pp. 262–276. Springer, Heidelberg (2004)
6. Benaloh, J., Chase, M., Horvitz, E., Lauter, K.E.: Patient controlled encryption: ensuring privacy of electronic medical records. In: Proceedings of the First ACM Cloud Computing Security Workshop, CCSW 2009, pp. 103–114 (2009)
7. Boneh, D., Di Crescenzo, G., Ostrovsky, R., Persiano, G.: Public key encryption with keyword search. In: Cachin, C., Camenisch, J.L. (eds.) EUROCRYPT 2004. LNCS, vol. 3027, pp. 506–522. Springer, Heidelberg (2004)
8. Boneh, D., Franklin, M.: Identity-based encryption from the weil pairing. In: Kilian, J. (ed.) CRYPTO 2001. LNCS, vol. 2139, pp. 213–229. Springer, Heidelberg (2001)
9. Boneh, D., Waters, B.: Conjunctive, subset, and range queries on encrypted data. IACR Cryptology ePrint Archive, p. 287 (2006)
10. Byun, J.W., Rhee, H.S., Park, H.-A., Lee, D.-H.: Off-line keyword guessing attacks on recent keyword search schemes over encrypted data. In: Jonker, W., Petković, M. (eds.) SDM 2006. LNCS, vol. 4165, pp. 75–83. Springer, Heidelberg (2006)

11. Cao, N., Wang, C., Li, M., Ren, K., and Lou, W.: Privacy-preserving multi-keyword ranked search over encrypted cloud data. In: 30th IEEE International Conference on Computer Communications INFOCOM 2011, pp. 829–837 (2011)
12. Chang, Y.-C., Mitzenmacher, M.: Privacy preserving keyword searches on remote encrypted data. In: Ioannidis, J., Keromytis, A.D., Yung, M. (eds.) ACNS 2005. LNCS, vol. 3531, pp. 442–455. Springer, Heidelberg (2005)
13. Di Crescenzo, G., Saraswat, V.: Public key encryption with searchable keywords based on jacobi symbols. In: Srinathan, K., Rangan, C.P., Yung, M. (eds.) INDOCRYPT 2007. LNCS, vol. 4859, pp. 282–296. Springer, Heidelberg (2007)
14. Curtmola, R., Garay, J.A., Kamara, S., Ostrovsky, R.: Searchable symmetric encryption: improved definitions and efficient constructions. IACR Cryptology ePrint Archive, p. 210 (2006)
15. Fang, L., Susilo, W., Ge, C., Wang, J.: Public key encryption with keyword search secure against keyword guessing attacks without random oracle. Inf. Sci. **238**, 221–241 (2013)
16. Goh, E.: Secure indexes. IACR Cryptology ePrint Archive, p. 216 (2003)
17. Golle, P., Staddon, J., Waters, B.: Secure conjunctive keyword search over encrypted data. In: Jakobsson, M., Yung, M., Zhou, J. (eds.) ACNS 2004. LNCS, vol. 3089, pp. 31–45. Springer, Heidelberg (2004)
18. Hwang, Y.-H., Lee, P.J.: Public key encryption with conjunctive keyword search and its extension to a multi-user system. In: Takagi, T., Okamoto, E., Okamoto, T., Okamoto, T. (eds.) Pairing 2007. LNCS, vol. 4575, pp. 2–22. Springer, Heidelberg (2007)
19. Jeong, I.R., Kwon, J.O., Hong, D., Lee, D.H.: Constructing PEKS schemes secure against keyword guessing attacks is possible? Comput. Commun. **32**(2), 394–396 (2009)
20. Li, M., Lou, W., Ren, K.: Data security and privacy in wireless body area networks. IEEE Wireless Commun. **17**(1), 51–58 (2010)
21. Li, M., Yu, S., Ren, K., Lou, W.: Securing personal health records in cloud computing: patient-centric and fine-grained data access control in multi-owner settings. In: Jajodia, S., Zhou, J. (eds.) SecureComm 2010. LNICST, vol. 50, pp. 89–106. Springer, Heidelberg (2010)
22. Liu, Q., Wang, G., Wu, J.: Secure and privacy preserving keyword searching for cloud storage services. J. Netw. Comput. Appl. **35**(3), 927–933 (2012)
23. Park, D.J., Kim, K., Lee, P.J.: Public key encryption with conjunctive field keyword search. In: Lim, C.H., Yung, M. (eds.) WISA 2004. LNCS, vol. 3325, pp. 73–86. Springer, Heidelberg (2005)
24. Rhee, H.S., Susilo, W., Kim, H.: Secure searchable public key encryption scheme against keyword guessing attacks. IEICE Electron Express **6**(5), 237–243 (2009)
25. Sun, W., Wang, B., Cao, N., Li, M., Lou, W., Hou, Y. T., Li, H.: Privacy-preserving multi-keyword text search in the cloud supporting similarity-based ranking. In: 8th ACM Symposium on Information, Computer and Communications Security, ASIA CCS, pp. 71–82. ACM (2013)
26. Swaminathan, A., Mao, Y., Su, G., Gou, H., Varna, A. L., He, S., Wu, M., Oard, D. W.: Confidentiality-preserving rank-ordered search. In: Proceedings of the ACM Workshop on Storage Security and Survivability, StorageSS, pp. 7–12 (2007)
27. van Liesdonk, P., Sedghi, S., Doumen, J., Hartel, P., Jonker, W.: Computationally efficient searchable symmetric encryption. In: Jonker, W., Petković, M. (eds.) SDM 2010. LNCS, vol. 6358, pp. 87–100. Springer, Heidelberg (2010)

28. Wang, C., Cao, N., Li, J., Ren, K., Lou, W.: Secure ranked keyword search over encrypted cloud data. In: International Conference on Distributed Computing Systems, ICDCS 2010, pp. 253–262. IEEE Computer Society (2010)
29. Wang, C., Cao, N., Ren, K., Lou, W.: Enabling secure and efficient ranked keyword search over outsourced cloud data. IEEE Trans. Parallel Distrib. Syst. **23**(8), 1467–1479 (2012)
30. Zhang, W. Lin, Y., Xiao, S., Liu, Q., Zhou, T.: Secure distributed keyword search in multiple clouds. In: IEEE 22nd International Symposium of Quality of Service, IWQoS 2014, pp. 370–379. IEEE (2014)

A Signature Generation Approach Based on Clustering for Polymorphic Worm

Jie Wang$^{(\boxtimes)}$ and Xiaoxian He

Central South University, Changsha, Hunan Province, China
jwang@csu.edu.cn

Abstract. To prevent worms from propagating rapidly, it is essential to generate worm signatures quickly and accurately. However, existing methods for generating worm signatures either cannot handle noise well or assume there is only one kind of worm sequence in the suspicious flow pool. We propose an approach based on seed extending signature generation (SESG) to generate polymorphic worm signatures from a suspicious flow pool which includes several kinds of worm and noise sequences. The proposed SESG algorithm computes the weight of every sequence, the sequences are queued based on their weight, and then classified. Worm signatures are then generated from the classified worm sequences. We compare SESG with other approaches. SESG can classify worm and noise sequences from a suspicious flow pool, and generate effective worm signatures more easily.

Keywords: Signature generation · Worm detection · Seed-extending algorithm · Polymorphic worm

1 Introduction

Worms are self-replicating malicious programs and represent a major security threat for the internet. They can infect and damage a large number of vulnerable hosts at timescales where human responses are unlikely to be effective [1]. According to an empirical study, a typical zero-day attack may last for 312 days on average [2]. Polymorphic worms allow a worm to change its appearance with every instance. They have caused great damage to the internet in recent years. Detecting and defending against polymorphic worm remains largely an open problem [3–5]. Most recent polymorphic worm detection research concentrates on signature-based detection [6–8]. These techniques look for specific byte sequences (called attack signatures) that are known to appear in the attack traffic. Their efficiency of defending against worms depends on the quality of worm signatures that can be generated. To detect polymorphic worms efficiently, accurate worm signatures must be first generated.

Existing approaches for automatically generating worm signatures include systems based on:

© Springer International Publishing Switzerland 2016
M. Yung et al. (Eds.): INTRUST 2015, LNCS 9565, pp. 84–96, 2016.
DOI: 10.1007/978-3-319-31550-8_6

1. Longest common string (LCS). Cai et al. [9] developed the WormShield system, Ranjan et al. [10] developed the DoWicher system, and Portokalidis et al. [11] developed the SweetBait system;
2. Semantics-aware. Yegneswaran et al. [12] proposed an architecture for generating semantics-aware signatures.
3. Common substrings (tokens). Newsome et al. [13] developed the Polygraph system which generates conjunction signatures, token-subsequence signatures, and Bayes signatures based on tokens extracted. Li et al. [14] developed the Hamsa system, which is an improvement over the Polygraph system in terms of both speed and attack resilience, but takes the number of substring token occurrence into consideration as part of the signature. Cavallaro et al. [15] proposed LISABETH, an improved version of Hamsa, an automated content-based signature generation system for polymorphic worms that uses invariant bytes analysis of network traffic content. Bayogle et al. [16] proposed Token-Pair Conjunction and Token-Pair Subsequence signature for detecting polymorphic worm threats. Wang [17] proposed an automated signature generation approach for polymorphic worms based on color coding.
4. Multi-sequence. Tang et al. [18] used multiple sequence alignment techniques to generate simplified regular expression (SRE) signatures for polymorphic worms.
5. Character frequency. Tang et al. [19] proposed an automated approach based on position-aware distribution signature (PADS) to defend against polymorphic worms.

Only Wang [17] and Tang et al. [18] discuss how to address noise in the process of generating a worm signature, but they both suppose that the suspicious pool only includes a single type of polymorphic worm. Tang et al. proposed an approach based on normalized cuts to cluster polymorphic worms, then generated a PADS signature for every cluster worm. However, in presence of noise, the accuracy of PADS cannot be assured [14]. The approaches discussed above can generate signatures for worms without noise and with one type of polymorphic worm in the suspicious pool, but they have difficulty generating worm signatures in the presence of noise and/or many type of polymorphic worms in the suspicious flow pool.

We propose a novel algorithm based seed extending signature generation (SESG) to generate polymorphic worm signatures from a suspicious flow pool which includes several types of worm and noise sequences. The SESG algorithm calculates the weight of every identified sequence, sequences are placed into a queue based on their weight, and all sequences are classified. Worm signatures are generated from the classified worm sequences. We compare SESG with other approaches, and show that SESG can classify worm sequences and noise sequences from suspicious flow pool, with lower false positive and false negative outcomes than current approaches.

Section 2 introduces a new worm signature, neighborhood-relation signature, which is used to detect polymorphic worms. Automatic polymorphic worm signature generation algorithm is proposed in Sect. 3. Experimental results are shown and discussed in Sect. 4. In Sect. 5 we present our conclusions.

2   Neighborhood-Relation Signatures

For a sequence $S = c_1c_2 \ldots c_m$, c_{i+1} is defined as neighbor of c_i, and $d_{i,i+1} = |c_{i+1} - c_i|$ is defined as the byte distance between c_i and c_{i+1}.

There is at least a significant region which infects the victim in polymorphic worms. The neighborhood-relation signatures (NRS) is generated by computing the byte distance of each position in the significant region, and it has a byte distance frequency distribution for each position in the signature string.

We first describe the process of computing an NRS from worm samples, then explain how to match a byte sequence against a signature.

2.1   Concept and Definition

Consider a set of worm sequences, $S = \{S_1, S_2, \ldots, S_n\}$, where $S_i = c_1c_2 \ldots c_m$. Suppose the starting positions of the significant region in n sequences are $a_1, a_2, \ldots, a_n$, and the width of a significant region is w.

The $count(p, d)$ is the number of worm sequences in which the neighbor distance of position, p, in the significant regions is d, where $d \in [0 \ldots 255]$. The neighbor distance distribution is

$$f_p(d) = \frac{count(p, d)}{n} , \qquad (1)$$

where $\sum_{d \in [0 \ldots 255]} f_p(d) = 1$, and $p = 1, 2, \ldots, w - 1$. $f_p(d)$ will be zero if $count(p, d) = 0$, which will cause signature generation to be undefined. Therefore, we set $f_p(d) = b$ for this case, where b is a small real number.

The NRS signature of n sequences is defined as $(f_1, f_2, \ldots, f_{w-1})$, where the width of the signature is $w - 1$.

2.2   Matching Between Sequences and NRS Signatures

Assume that S_i is a byte sequence and l is the length of S_i. $d_{1,2}, d_{2,3}, \ldots, d_{l-1,l}$ denotes the 1-neighbor distances of position $1, 2, \ldots, l - 1$ in S_i respectively. Let $seg(S_i, a_i)$ be the substring of S_i with starting position a_i and width w. The matching score of $seg(S_i, a_i)$ is

$$Score = \Pi_{p=1}^{w-1} \frac{f_p(d_{a_i+p,a_i+p+1})}{f} ,$$

where $f = \frac{1}{255}$.

There exists a position a_i of S_i that maximizes $Score$. The matching score of sequence S_i is defined as

$$maxScore = \max_{a_i=1}^{l-w+1} \Pi_{p=1}^{w-1} \frac{f_p(d_{a_i+p,a_i+p+1})}{f} .$$

For ease of plotting, we use the logarithm of $maxScore$ as the final matching score of S_i with the NRS signature, that is

$$\Theta = \max_{a_i=1}^{l-w+1} \frac{1}{w-1} \sum_{p=1}^{w-1} log \frac{f_p(d_{a_i+p,a_i+p+1})}{f} . \tag{2}$$

The w-byte segment is considered as the significant region if its matching score is equal to Θ. The significant region is the segment of S_i that matches best with the NRS signature. When $\Theta > 0$, S_i is classified as a worm sequence.

2.3   NRS Generation Algorithm

We have shown that the NRS and the significant regions can lead to each other. We now explain how to generate NRS using Gibbs sampling if we do not know either the NRS signature or significant regions.

The Gibbs algorithm randomly assigns starting positions $a_1, a_2, \ldots, a_n$ of the significant regions for a set S of worm variants, $S_1, S_2, \ldots, S_n$. If S_1 of S is chosen, then $f_p(d)$ as defined in Eq. (1) is calculated based on the significant regions for the other $S - S_1$ variants, and the NRS is calculated. The new estimated position for the significant region for S_1 is calculated based on the estimated NRS, and the starting position a_1 is modified accordingly. The remaining members S_2, $S_3, \ldots, S_n$ of S are chosen in turn until the termination condition is satisfied. After S_n is chosen, if the termination condition is not satisfied, the sampling algorithm continues to run, choosing S_1 again.

The Gibbs sampling algorithm terminates if the average matching score between the worm variants and the signature is within $(1 \pm \varepsilon)$ of the average matching score of the previous t iterations, where ε is a small predefined percentage. The Gibbs sampling algorithm is illustrated in Fig. 1.

3   SESG Approach for Polymorphic Worms

The suspicious flow pool is represented as an undirected complete graph $G(V, E)$ with sequences as vertices. We propose an SESG approach for polymorphic worms. SESG can be divided into four major parts: Weighting Vertex, Selecting Seed, Extending Cluster, and Generating Signature. The input to the algorithm is an undirected complete graph.

3.1   Weighting Vertex

An approach based on NRS against polymorphic worms was proposed by Wang et al. [20]. We apply NRS based on 1-neighbor distances and the Gibbs sampling algorithm. Let W_i be the weight of vertex S_i, and Θ_{ij} the weight of an edge $[S_i, S_j]$. The NRS is generated from S_i and S_j by applying the Gibbs sampling algorithm. Then the matching score $\Theta(NRS, S_i)$ between NRS and S_i and the matching score $\Theta(NRS, S_j)$ between NRS and S_j are calculated using the

The Gibbs(S, w) algorithm
Input: n sequences $S = S_1, S_2, \ldots, S_n$, the width of significant regions w;
Output: NRS;
Width=w;
Randomly assigning the starting positions $a_1, a_2, \ldots, a_n$ of the significant regions for sequences $S_1, S_2, \ldots, S_n$;
j=1;
For $i = 1$ to 5000
    {$j = (j \mod n)$;
    {NRS=SigGen($S - S_j$,$a_1, a_2, \ldots, a_{j-1}, a_{j+1}, \ldots, a_n, w$);
    $a_j = Updateloc(S_j, NRS)$;
    $j + +$;
    For $l = 1$ to n
        $maxScore_l \leftarrow ComScore(S_l, NRS)$);
    $Ascore_i = \frac{1}{n} \sum_{l=1}^{n} maxScore_l$;
    If $|Ascore_i - \frac{1}{t} \sum_{l=i-t}^{i-1} Ascore_l| < \varepsilon$ then break;}
Return NRS;

Fig. 1. The Gibbs sampling algorithm

matching process introduced in Wang et al. [20]. The weight of an edge is then defined as

$$\Theta_{ij} = \Theta(NRS, S_i) + \Theta(NRS, S_j) , \qquad (3)$$

where Θ_{ii} is set to 0. We define the weight of each vertex to be the sum of the weights of its adjacent edges, $W_i = \sum_j \Theta_{ij}$. After all vertices are assigned weights, we sort the vertices in non-increasing order by their weights and store them in a queue, Q.

3.2   Selecting the Seed Vertex

We pick the first (highest weight) vertex in Q and use it as a seed to grow a new cluster. Once the cluster is completed, all vertices in the cluster are removed from Q, and we pick the first remaining vertex in Q as the seed for the next cluster.

3.3   Extending a Cluster

A cluster, K, is extended by adding vertices recursively from its neighbors according to their priority. A neighbor is added to the cluster if the weight of its adjacent edge with the seed is higher than existing members of the cluster. We calculate the NRS from two vertices of the cluster and calculate the matching scores between NRS and the neighbors. The matching scores are considered as the priorities of the neighbors. If the highest priority is greater than zero, the neighbor with the highest priority will be added to the cluster. Once the new vertex is added to the cluster, the cluster is updated, i.e., the neighbors of the new cluster are re-constructed and the priorities of the neighbors of the new cluster are re-calculated, and the algorithm proceeds recursively with the new cluster until all the priorities are less than zero.

3.4   Generating the Signature

When Q is empty, all vertices are divided into several clusters. We generate NRS from every cluster. These NRS are filtered in a normal flow pool including m sequences. The matching scores between a NRS and m sequences are $\Theta_1, \Theta_2, \ldots, \Theta_m$. If $\Theta_i > 0(1 \leq i \leq m)$, p increments by 1. If $p/m < \varepsilon$, the NRS is considered to be the signature of a worm sequence, where ε is a small predefined percentage. Otherwise the NRS will be filtered out.

The SESG process is illustrated in Fig. 2.

The SESG(N, M) algorithm
Input: n sequences $N = x_1, x_2, \ldots, x_n$; m normal sequences $M = M_1, M_2, \ldots, M_m$;
Output: y sets of worm sequences $S_1, S_2, \ldots, S_y$ and their NRS;
(**Weighting Vertex**)
$NRS \leftarrow \text{Gibbs}(x_i, x_j)$;
$\Theta_{ij} = \Theta(NRS, X_i) + \Theta(NRS, x_j)$;
Constructing Q based on sequence weight W;
(**Selecting seed**)
While Q is not empty
   $x \leftarrow$ the first sequence in Q; $S = x$;
   call Extending Cluster(S);
   $Q \leftarrow Q - S$;
(**Extending Cluster(S)**)
For i=1 to n
   If x_i does not belong to any cluster then $Q_S \leftarrow \Theta_{xx_i}$;
   If Θ_{xv} is the largest value in the set Q_S
      Cluster$(S) \leftarrow v$;
$NRS \leftarrow \text{Gibbs}(S)$;
For i=1 to n
   {If sequence x_i does not belong to any cluster, then
      {Calculate the match score $Score_i$;
      If all match scores are less than 0 then break;
      else Cluster$(S) \leftarrow x$;}
(**Generating Signature**)
For each cluster
   {$NRS \leftarrow \text{Gibbs}(\text{Cluster})$;
   Filtering NRS in M and getting the last Cluster;
Return (Cluster, NRS);

Fig. 2. The SESG Algorithm

4   Experiments and Results

We tested the accuracy and effectiveness of the SESG algorithm with and without noise in the environment, and compared with other algorithms. Four kinds of worms are used in the experiments: MS Blaster, SQL Slammer, Apache-Knacker, and ATPhttp. We apply polymorphism techniques, including encryption, garbage-code insertion, instruction rearrangement, and substitution, to generate polymorphic samples for these worms.

4.1   Experiment 1: Noise and a Single Worm Sequence

The suspicious flow pool includes 50 noise sequences and 50 Blaster worm sequences. We apply the SESG algorithm and the classification method based on normalized cuts (NC) [19] to classify sequences of the suspicious pool. The NC algorithm needs to know the number of classifications prior to running, and so we first set the number of classifications = 2. Classification results of SESG are illustrated in Fig. 3, for NC in Fig. 4. In both figures, the x-axis 1–50 are the Blaster worm and x-axis 51–100 are noise sequences.

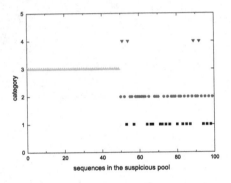
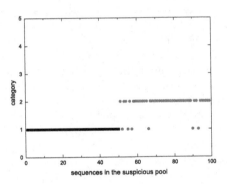

Fig. 3. Experiment 1: Classification from SESG

Fig. 4. Experiment 1: Classification from NC

The SESG algorithm successfully distinguished worms and noise sequences, dividing the 100 sequences into 4 groups. The Blaster worms were all allocated into the third. Different noise sequences have different characteristics, so the noise sequences were divided into 3 classes. On the other hand, the NC algorithm allocated the Blaster worms and 8 noise sequences into the first class.

We generated NRS from the classified worm sequences, and used 10000 Blaster worms and 10000 normal flow sequences to test the false negative and false positive ratio. We applied the Gibbs algorithm [20] (GNRS), Polygraph [13], and CCSF [17] (CNRS) to generate NRS from the unclassified suspicious flow pool. Because of the noise sequences in the suspicious flow pool, Polygraph could not generate worm signatures. The test results are shown in Table 1.

Table 1. Experiment 1: Blaster worm

	GNRS	CNRS	NC+NRS	SESG+NRS
False positive ratio	0.4315	0	0.1017	0
False negative ratio	0	0	0	0

The false positive and false negative ratios of NRS generated by SESG are 0. SESG successfully distinguishes worms and noise sequences, and NRS is generated only from Blaster samples. Therefore, the NRS has no false negatives or positives. However, NC generates NRS from sample sequences including 8 noise sequences. Hence NRS has false positive 0.1017. GNRS is generated from sample sequences including 50 noise sequences, so it has higher false positive. The CCSF algorithm can remove noise, so both the false positive and false negative ratios are 0.

4.2   Experiment 2: Two Worm Sequences

The suspicious flow pool includes 50 SQL Slammer worm sequences and 50 Blaster worm sequences. Classification results from the SESG algorithm are shown in Fig. 5, and for the NC algorithm in Fig. 6. In both figures x-axis 1–50 are Blaster worms and x-axis 51–100 are SQL Slammer worms.

The SESG algorithm divided the 100 sequences into 2 groups, with Blaster worms in one class and SQL Slammer worms in the other. The NC algorithm also divided the 100 sequences into 2 groups. However, while the second class includes only Blaster worms, the first class includes 50 SQL Slammer worms and 2 Blaster worms.

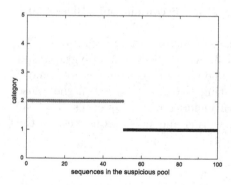
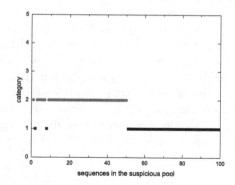

Fig. 5. Experiment 2: classification from SESG

Fig. 6. Experiment 2: classification from NC

Just as Sect. 4.1, we applied different algorithms to generate worm signatures. Because the suspicious flow pool includes two types of worms, Polygraph and CCSF cannot generate NRS signatures. Both the false positive and false negative ratios of the SESG and NC algorithms are 0.

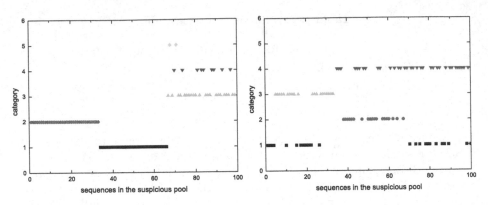

Fig. 7. Experiment 3: classification from SESG

Fig. 8. Experiment 3: classification from NC

4.3 Experiment 3: Noise and Two Types of Worm Sequences

The suspicious flow pool includes 33 noise sequences, 33 SQL Slammer worm sequences, and 34 Blaster worm sequences. Classification results for the SESG algorithm are shown in Fig. 7, and for the NC algorithm in Fig. 8. In both figure, x-axis 1–34 Blaster worms, 35–67 are SQL Slammer worms, and 68–100 are noise sequences.

The SESG algorithm divided 100 sequences into 5 groups. SQL Slammer worms were divided into the first class, Blaster worms into the second class, and the other three classes contain only noise sequences. Whereas the NC algorithm divides the 100 sequences into 4 groups. Some noise sequences and Blaster worms are allocated to the first group, some noise sequences and SQL Slammer worms were allocated to the fourth group. Group 2 and 3 contain only Blaster and Slammer worms respectively.

As above, we apply different algorithms to generate worm signatures. The test results are illustrated in Tables 2 and 3. As for Sect. 4.2, Polygraph and CCSF could not generate NRS signatures. The Gibbs algorithm cannot remove noise disturbance in the process of generating NRS, and so has higher false positive ratio when the suspicious flow pool includes noise sequences. The NC algorithm does not completely distinguish worms and noise sequences, so the NRS generated by NC has higher positive ratio.

Table 2. Experiment 3: Blaster worm

	GNRS	NC+NRS (the first class)	NC+NRS (the third class)	SESG+NRS
False positive ratio	0	0	0	0
False negative ratio	0.3266	0.4221	0	0

Table 3. Experiment 3: SQL Slammer worm

	GNRS	NC+NRS (the second class)	NC+NRS (the fourth class)	SESG+NRS
False positive ratio	0	0	0	0
False negative ratio	0.3859	0	0.4349	0

4.4 Experiment 4: Four Worm Sequences

The suspicious flow pool includes four types of worm sequences, 25 of each type. Classification results from the SESG algorithm are shown in Fig. 9, and from the NC algorithm in Fig. 10. In both figures x-axis 1–25 are Blaster worms, 25–50 are SQL Slammer worms, 51–75 are Apache-Knacker worms, and 76–100 ATPhttp worms.

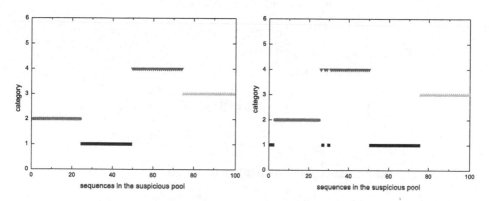

Fig. 9. Experiment 4: classification from SESG

Fig. 10. Experiment 4: classification from NC

The SESG divided the 100 sequences into 4 groups, successfully distinguishing the different worms. The NC algorithm also divided the 100 sequences into 4 groups. However, the groups include several worm types. For example, the first group includes 2 SQL Slammer worms, 2 Blaster worms and 25 Apache-Knacker worms.

As above, we apply different algorithms to generate worm signatures. Since the suspicious flow pool includes four kinds of worms, Polygraph and CCSF cannot generate NRS signatures. The false positive and false negative ratios of all the algorithms are 0, since there was no noise included.

4.5 Experiment 5: Noise Sequences and Four Types of Worm Sequences

The suspicious flow pool includes noise sequences and four types of worm sequences, 20 of each class. Classification results from the SESG algorithm are

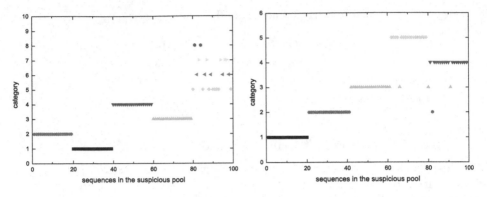

Fig. 11. Experiment 5: classification from SESG

Fig. 12. Experiment 5: classification from NC

Table 4. Experiment 5: Blaster worm

	GNRS	NC+NRS	SESG+NRS
False positive ratio	0	0	0
False negative ratio	0.3146	0	0

Table 5. Experiment 5: SQL Slammer worm

	GNRS	NC+NRS	SESG+NRS
False positive ratio	0	0	0
False negative ratio	0.2763	0.047	0

shown in Fig. 11, and for the NC algorithm in Fig. 12. In both figures, x-axis 1–20 are Blaster worms, 20–40 are SQL Slammer worms, 41–60 are Apache-Knacker worms, 61–80 are ATPhttp, and 81–100 are noise sequences.

The SESG algorithm divided the 100 sequences into 8 groups. SQL Slammer worms are allocated to the first class, Blaster worms into the second, ATPhttp worms the third, and Apache-Knacker worms into the fourth class. The remaining four classes are noise sequences. The NC algorithm divides the 100 sequences into 5 groups. However, some noise sequences and some worm sequences are allocated into the same group. For example, the second group includes 20 SQL Slammer worms, 1 Apache-Knacker worms, and 2 noise sequences; and the third group includes 2 noise sequences, 3 ATPhttp worms, and 20 Apache-Knacher worms. The NC algorithm classifies sequences step by step, and the number of classifications is set beforehand. The NC algorithm first divides the sequences of the suspicious flow pool into 2 groups, then continues to divide the classified sequences into 2 subclasses, etc. Previous classification results usually influence later classifications. Therefore, the NC algorithm cannot distinguish worms and noise sequences.

As above, we applied different algorithms to generate worm signatures, and the test results are shown in Tables 4, 5, 6 and 7 for the four worms Blaster, SQL

Table 6. Experiment 5: Apache-Knacker worm

	GNRS	NC+NRS	SESG+NRS
False positive ratio	0	0	0
False negative ratio	0.3276	0.076	0

Table 7. Experiment 5: ATPhttp worm

	GNRS	NC+NRS	SESG+NRS
False positive ratio	0	0	0
False negative ratio	0.3257	0	0

Slammer, ATPhttp, and Apache-Knacker, respectively. Since the suspicious flow pool includes four kinds of worms and noise sequences, polygraph and CCSF cannot generate NRS signatures. For all worms, NRS generated by the Gibbs and NC algorithms have high false positive ratios. The false positive and false negative ratios of the SESG algorithm are both 0.

5   Conclusion

We propose an approach based on SESG to generate accurate polymorphic worm signatures from a suspicious flow pool that includes several types of worm and noise sequences. The SESG method consists of four subprograms: Weighting Vertex, Selecting Seed, Extending Cluster and Generating Signature. We showed that, compared with other approaches, SESG can accurately classify worm and noise sequences from a suspicious flow pool where there are multiple signatures and noise present. Combined with NRS, SESG can act as a single polymorphic worm signature generation algorithm, and can also be used as a classification algorithm with significantly better accuracy than current algorithms, such as CCSF and Polygraph, etc.

Acknowledgments. This work is supported by National Natural Science Foundation of China under Grant No.61202495.

References

1. Antonatos, S., Akritidis, P., Markatos, E.P., Anagnostakis, K.G.: Defending against hitlist worms using network address space randomization. Comput. Netw. **51**(12), 3471–3490 (2007)
2. Bilge, L., Dumitras, T.: Before we knew it: an empirical study of zero-day attacks in the real world. In: Proceedings of ACM Conference on Computer and Communications Security (CCS 2012), New Carolina, pp. 833–844, October 2012
3. Sun, W.C., Chen, Y.M.: A rough set approach for automatic key attributes indentification of zero-day polymorphic worms. Expert Syst. Appl. **36**(3), 4672–4679 (2009)

4. Mohammed, M.M.Z.E., Chan, H.A., Ventura, N., Hashim, M., Bashier, E.: Fast and accurate detection for polymorphic worms. In: Proceedings of Internetional Conference for Internet Technology and Secured Transactions, pp. 1–6 (2010)
5. Comar, P.M., Liu, L., Saha, S., Tan, P.N., Nucci, A.: Combining supervised and unsupervised learning for zero-day malware detection. In: Proceedings of 32nd Annual IEEE International Conference on Computer Communications (INFOCOM 2013), Turin, Italy, pp. 2022–2030, April 2013
6. Bayoglu, B., Sogukpinar, L.: Graph based signature classes for detecting polymorphic worms via content analysis. Comput. Netw. **56**(2), 832–844 (2012)
7. Tang, Y., Xiao, B., Lu, X.: Signature tree generation for polymorphic worms. IEEE Trans. Comput. **60**(4), 565–579 (2011)
8. Modi, C., Patel, D., Borisaniya, B., Patel, H., Patel, A., Rajarajan, M.: A survey of intrusion detection techniques in cloud. J. Netw. Comput. Appl. **36**(1), 42–57 (2013)
9. Cai, M., Hwang, K., Pan, J., Christos, P.: WormShield: fast worm signature generation with distributed fingerprint aggregation. IEEE Trans. Dependable Secure Comput. **5**(2), 88–104 (2007)
10. Ranjan, S., Shah, S., Nucci, A., Munafo, M., Cruz, R., Muthukrishnan, S.: DoWitcher: effective worm detection and containment in the internet core. In: IEEE Infocom, Anchorage, Alaskapp, pp. 2541–2545 (2007)
11. Portokalidis, G., Bos, H.: SweetBait: zero-hour worm detection and containment using low- and high-interaction honeypots. Comput. Netw. **51**(11), 1256–1274 (2007)
12. Yegneswaran, V., et al.: An architecture for generating semantics-aware signatures. In: Proceedings of the 14th conference on USENIX Security Symposium. USENIX Association, Berkeley (2005)
13. Newsome, J., Karp, B., Song, D.: Polygraph: automatically generation signatures for polymorphic worms. In: Proceedings of 2005 IEEE Symposium on Security and Privacy Symposium, Oakland, pp. 226–241 (2005)
14. Li, Z., Sanghi, M., Chen, Y., Kao, M., Chavez, B.: Hamsa: fast signature generation for zero-day polymorphic worms with provable attack resilience. In: Proceedings of IEEE Symposium on Security and Privacy, Washington, DC, pp. 32–47 (2006)
15. Cavallaro, L., Lanzi, A., Mayer, L., Monga, M.: LISABETH: automatedcontent-based signature generator for zero-day polymorphic worms. In: Proceedings of the Fourth International Workshop on Software Engineering for Secure Systems, Leipzig, pp. 41–48 (2008)
16. Bayoglu, B., Sogukpinar, L.: Polymorphic worm detection using token-pair signatures. In: Proceedings of the 4th International Workshop on Security, Privacy and Trust in Pervasive and Ubiquitous Computing, Sorrento, Italy, pp. 7–12 (2008)
17. Wang, J., Wang, J.X., Chen, J.E., Zhang, X.: An automated signature generation approach for polymorphic worm based on color coding. J. Softw. **21**(10), 2599–2609 (2010)
18. Tang, Y., Xiao, B., Lu, X.: Using a bioinformatics approach to generate accurate exploit-based signatures for polymorphic worms. Comput. Secur. **288**, 827–842 (2009)
19. Tang, Y., Chen, S.: An automated signature-based approach against polymorphic internet worms. IEEE Trans. Parallel Distrib. Syst. **18**, 879–892 (2007)
20. Wang, J., Wang, J.X., Sheng, Y., Chen, J.E.: Novel approach based on neighborhood relation signature against polymorphic internet worms. J. Commun. **32**(8), 150–158 (2011)

Security Model

Computational Soundness of Uniformity Properties for Multi-party Computation Based on LSSS

Hui Zhao[1,2(✉)] and Kouichi Sakurai[1]

[1] Kyushu University, 744 Motooka, Nishi-ku, Fukuoka 819-0395, Japan
eric.hui.zhao@gmail.com
[2] Shandong University of Technology, Zibo 255000, Shandong, China

Abstract. We provide a symbolic model for multi-party computation based on linear secret-sharing scheme, and prove that this model is computationally sound: if there is an attack in the computational world, then there is an attack in the symbolic (abstract) model. Our original contribution is that we deal with the uniformity properties, which cannot be described using a single execution trace, while considering an unbounded number of sessions of the protocols in the presence of active and adaptive adversaries.

Keywords: Multi-party computation · Uniformity properties · Universally composable

1 Introduction

1.1 Background and Motivation

Provable security are now widely considered an essential tool for validating the design of cryptographic schemes. While Dolev-Yao models traditionally comprise only non-interactive cryptographic operations, recent cryptographic protocols rely on more sophisticated interactive primitives, with unique features that go far beyond the traditional goals of cryptography to solely offer secrecy and authenticity of communication.

Secret-sharing cryptographic operations constitutes arguably one of the most prominent and most amazing such primitive [1–9]. Traditionally, in an linear secret-sharing scheme with threshold (t, l), a dealer D and a number of players $P_1, P_2, ..., P_l$ wish to securely generate the secret share $ss_1, ss_2, ..., ss_l$, where each player P_i holds a private input d_i. This secret-sharing scheme is considered sound and complete if any t or more valid secret shares make the secret s computable, and knowledge of any $t - 1$ or fewer secret shares leaves the secret completely undetermined.

Indeed, verifiable secret sharing (VSS) and secure multi-party computation (MPC) among a set of n players can efficiently be based on any linear secret-sharing scheme (LSSS) for the players, provided that the access structure of the

© Springer International Publishing Switzerland 2016
M. Yung et al. (Eds.): INTRUST 2015, LNCS 9565, pp. 99–113, 2016.
DOI: 10.1007/978-3-319-31550-8_7

LSSS allows MPC or VSS at all [9,18]. Secure multi-party computation (MPC) can be defined as the problem of n players to compute an agreed function of their inputs in a secure way, where security means guaranteeing the correctness of the output as well as the privacy of the players' inputs, even when some players cheat. A key tool for secure MPC, interesting in its own right, is verifiable secret sharing (VSS): a dealer distributes a secret value s among the players, where the dealer and/or some of the players may be cheating. It is guaranteed that if the dealer is honest, then the cheaters obtain no information about s, and all honest players are later able to reconstruct s. Even if the dealer cheats, a unique such value s will be determined and is reconstructible without the cheaters' help.

Thus, it is important to develop computationally sound abstraction techniques to reason about LSSS-based MPC (LMPC) and to offer support for the automated verification of their security.

1.2   Related Works

Starting with the seminal work of Abadi and Rogaway [10–12], a lot of efforts has been directed to bridging the gap between the formal analysis system and computational-soundness model. The goal is to obtain the best of both worlds: simple, automated security proofs that entail strong security guarantees. Research over the past decade has shown that many of these Dolev-Yao models are computationally sound, i.e., the absence of attacks against the symbolic abstraction entails the security of suitable cryptographic realizations. Most of these computational soundness results against active attacks, however, have been specific to the class of trace properties, which is only sufficient as long as strong notions of privacy are not considered, e.g., in particular for establishing various authentication properties [13–18]. Only few computational soundness results are known for the class of equivalence properties against active attackers, most of these results focus on abstractions for which it is not clear how to formalize any equivalence property beyond the non-interactive cryptographic operations [19–22], such as multi-party computation that rely on more sophisticated interactive primitives [13,18].

1.3   Challenging Issues

Canetti et al. proposed framework of universally composable security [18,21]. In this framework, security properties is defined by what it means for a protocol to realize a given ideal functionality, where an ideal functionality is a natural algorithmic way of capturing the desired functionality of the protocol problem at hand. A protocol that is secure within the universally composable framework is called universally composable (UC).

We are facing a situation where computational soundness results, despite tremendous progress in the last decade, still fall short in comprehensively addressing the class of equivalence properties and protocols that formal verification tools are capable to deal with. Moreover, it is still unknown the composability in UC framework can be extended to achieve more comprehensive computational soundness results for equivalence properties.

1.4   Our Main Contributions

In this paper, we present an abstraction of LSSS within the pi-calculus. In this abstraction, linear secret-sharing scheme is defined by equational theory. Further, abstraction of LMPC is provided as a process that receives the inputs from the parties involved in the protocol over private channels, create secret shares of the MPC result, and sends secret shares to the parties again over private channels. This abstraction can be used to model and reason about larger MPC protocols that employ LSSS as a building block.

We establish computational soundness results of preservation of uniformity properties for protocols built upon our abstraction of LMPC. This result is obtained in essentially two steps: We first establish an ideal functionality for LMPC in the UC framework. Second, we obtain an secure cryptographic realization of our symbolic abstraction of LMPC. This computational soundness result holds for LMPC that involve arbitrary arithmetic operations; moreover, we will show that it is compositional, in the sense that uniformity properties of bi-processes in pi-calculus implies computational soundness results of preservation of uniformity properties. Such a result allows for soundly modeling and verifying many applications employing LMPC as a building block.

1.5   Comparison with Existing Works

While computational soundness proofs for Dolev-Yao abstractions of multi-party computation use standard techniques [13, 18, 23–25] finding a sound ideal functionality for multi-party computation do not support equivalence properties. Securely realizable ideal functionalities constitute a useful tool for proving computational soundness for equivalence properties of a Dolev-Yao model. In our proof, we establish a connection between our symbolic abstraction of LMPC and such an ideal functionality which supports uniformity properties: for an expressive class of equivalence properties, the uniformity implies observational equivalence for bi-processes, which are pairs of processes that differ only in the messages they operate on but not in their structure. We exploit that this ideal functionality is securely realizable.

2   The Abstraction of Secret-Sharing in Multi-party Computation

In this section, we present a symbolic abstraction of LMPC based on Backes's abstraction for MPC [13]. Since the overall protocol may involve several secure LMPC, a session identifier sid is often used to link the private inputs to the intended session. We then represent $SS_{l,t}(m, r)$ as a function that explicitly generates the LSSS-based proofs. The resulting abstraction of LMPC is depicted as the pi-calculus process $LMPC$ as follows.

$$Input_i = !(inloop_i(z).in_i(d_i, sid').\overline{adv}(sid'). \text{ if } sid = sid'$$
$$\text{then } \overline{lin}_i(d_i) \text{ else } \overline{inloop_i}(sync())) \mid \overline{inloop_i}(sync())$$

$$Deliver_i = \overline{in}_i(y_i, sid).\overline{inloop_i}(sync())$$

$$SSCompu(l, t, F)$$
$$= lin_1(d_1).lin_2(d_2)...lin_l(d_l).vr$$
$$\text{let } y = SS_{l,t}(F(d_1, d_2, ..., d_l), r) \text{ in}$$
$$\text{let } y_1 = COE_{l,t,1}(y) \text{ in}$$
$$\text{let } y_2 = COE_{l,t,2}(y) \text{ in}$$

$$\vdots$$

$$\text{let } y_l = COE_{l,t,l}(y) \text{ in}$$
$$(Deliver_1 \mid Deliver_2 \mid ... \mid Deliver_l)$$

$$LMPC(l, t, F, sidc, adv, \widetilde{in}) = sidc(sid).vlin.vinloop.$$
$$(Input_1 \mid Input_2 \mid ... \mid Input_l \mid SSCompu(l, t, F))$$

In the abstraction of LMPC above, process $LMPC$ is parametrized by threshold (l, t), l-function F, a session identifier channel $sidc$, a adversary channel adv, and l private channels in_i for the l players. Private channels in_i are authenticated such that only the ith player. The actual generation of secret share proof is performed in the last subprocess: after the private inputs of the individual parties are collected from the internal input channels lin_i, the function $SS_{l,t}(F(d_1, d_2, ..., d_l), r)$ is executed. After each computation round, the subprocesses $deliver_i$ send the individual secret-sharing proof $COE_{l,t,i}(SS_{l,t}(F(d_1, d_2, ..., d_l), r))$ over the private channels in_i to every player i along with the session identifier sid. In order to trigger the next round, $sync()$ is sent over the internal loop channels $inloop_i$.

The abstraction allows for a large class of l-function F as described below.

Definition 1 *(l-function). We call a term F an l-function if the term contains neither names nor variables, and if for every α_m occurring therein, we have $m \in [1, l]$.*

In our abstraction model for LMPC with (l, t) threshold, the values α_i in F constitute placeholders for the private input d_i. Based on that, Dealer and players' ability to produce LMPC proofs is modeled by introducing symbolic constructor $SS_{l,t}(F(\widetilde{d}), r)$, called *secret share key*. Its arguments are a message $F(\widetilde{d})$ and d_i will serve as substitutes for the variables α_i in F. The semantics of these constructors imply that the secret share proof indeed guarantees that in the abstract model the soundness and the completeness of the secret-sharing schemes with threshold (l, t). In the symbolic secret share proof above, $d_1, d_2, ..., d_l$ represent player $1, 2, ..., l$'s respectful private input.

In the following we define the symbolic model in which the execution of a symbolic protocol involving secret share proofs takes place.

First, we fix several countably infinite sets. By *Nonce* we denote the set of nonces. We use elements from *Garbage* to represent ill formed messages (corresponding to unparseable bitstrings in the computational model). Finally, elements of *Rand* denote symbolic randomness used in the construction of secret share proofs. We assume that *Nonce* is partitioned into in infinite sets $Nonce_{ag}$ and $Nonce_{adv}$, representing the nonces of honest agents and the nonces of the adversary. Similarly, *Rand* is partitioned into in infinite sets $Rand_{ag}$ and $Rand_{adv}$.

We proceed by defining the syntax of messages that can be sent in a protocol execution. Since such messages can contain secret share proofs, and these are parametrized over constructors that are to be computed, we first have to define the syntax of these constructors. Let the message type T be defined by the following grammar:

$$T ::= COE_{l,t}(SS_{l,t}(T,N)) \mid SS_{l,t}(T,N) \mid$$
$$pair(T,T) \mid S \mid N \mid garbage(N)$$
$$S ::= empty \mid string_0(S) \mid string_1(S)$$

The intuitive interpretation of a l-function is that it is a term with free variables α_i. The α_i will be substituted with messages.

We define destructors as follows:

$$D := \{Combin_{l,t}/t, fst/1, snd/1, unstring_0/1, unstring_1/1, equals/2\}.$$

The destructor $Combin_{l,t}$ extracts the secret from a secret share sequence. The destructors fst and snd are used to destruct pairs, and the destructors $unstring_0$ and $unstring_1$ allow to parse payload-strings.

We further define the equational rule as follows:

$$Combin_{l,t}(COE_{l,t,i_1}(SS_{l,t}(m,r)), COE_{l,t,i_2}(SS_{l,t}(m,r)), ...,$$
$$COE_{l,t,i_t}(SS_{l,t}(m,r)) = m.$$

Based on the symbolic model above, we then give the definition of symbolic pi-calculus execution of LMPC.

Definition 2 *(Symbolic pi-calculus execution). Let Π be a closed process, and let Adv be an interactive machine called the attacker. We define the symbolic pi-execution as an interactive machine $SExec_\Pi$ that interacts with Adv:*

1. *Start. Let $P = \Pi$, where we rename all bound variables and names (including nonces and randomness) such that they are pairwise distinct and distinct from all unbound ones. Let η and μ be a totally undefined partial functions from variables and names, respectively, to terms. Let $a_1, a_2, ..., a_n$ denote the free names in P_0. For each i, pick $r_i \in Nonces_{ag}$ at random. Set $\mu := \mu \bigcup \{a_1 := r_1, a_2 = r_2, ..., a_n := r_n\}$. Send $(r_1, r_2, ..., r_n)$ to Adv.*
2. *Main loop. Send P to the adversary and expect an evaluation context E from the adversary. Distinguish the following cases:*

(a) $P = E[M(x) : P_1]$: *Request two terms* c, m *from the adversary. If* $c = eval_{\eta,\mu}(M)$, *set* $\eta := \eta \bigcup \{x := m\}$ *and* $P := E[P_1]$.

(b) $P = E[va.P_1]$: *Pick* $r \in Nonce_{ag} \setminus range(\mu)$, *set* $P := E[P_1]$ *and* $\mu := \mu(a := r)$.

(c) $P = E[\widetilde{M}(N).P_1][M_2(x).P_2]$: *If* $eval_{\eta,\mu}(M_1) = eval_{\eta,\mu}(M_2)$, *then set* $P := E[P_1][P_2]$ *and* $\eta := \eta \bigcup \{x := eval_{\eta,\mu}(N)\}$.

(d) $P = E[let\ x = D\ in\ P_1\ else\ P_2]$: *If* $m := eval_{\eta,\mu}(D)\bot$, *set* $\eta := \eta \bigcup \{x := m\}$ *and* $P := E[P_1]$; *Otherwise set* $P := E[P_2]$.

(e) $P = E[!P_1]$: *Rename all bound variables of* P_1 *such that they are pairwise distinct and distinct from all variables and names in* P *and in the domains of* η *and* μ, *yielding a process* $\widetilde{P_1}$. *Set* $P := E[\widetilde{P_1}|!P_1]$.

(f) $P = E[\widetilde{M}(N).P_1]$: *Request a term* c *from the adversary. If* $c = eval_{\eta,\mu}(M)$, *set* $P := E[P_1]$ *and send* $eval_{\eta,\mu}(N)$ *to the adversary.*

(g) *In all other cases, do nothing.*

We are now ready to define what uniformity properties of a bi-protocol in the applied pi-calculus is.

Definition 3 *(Bi-process). A pi-calculus Bi-process* Π *is defined like a protocol but uses bi-terms instead of terms. A bi-term is a pair (left, right) of of two (not necessarily distinct) pi-calculus terms in the protocol definition. In the left process* $left(\Pi)$ *the bi-terms are replaced by their left components; the right process* $right(\Pi)$ *is defined analogously.*

Definition 4 *(Uniformity properties for bi-process). We say that the bi-process* Π *is uniform if for any interactive machine* Adv,
$$SExec_{left(\Pi),Adv} \approx SExec_{right(\Pi),Adv}.$$

3   Computational Soundness of Secret-Sharing in Multi-party Computation

In this section, we present a computational soundness result of preservation of uniformity properties for our abstraction of LMPC. Our result builds on the universally composable (UC) framework [18,21], where the security of a protocol is defined by comparison with an ideal functionality I. The proof proceeds in three steps, as depicted in the following.

In the first step, we prove that the uniformity properties of an applied pi-calculus process carries over to the computational setting, where the protocol is executed by interactive Turing machines operating on bitstrings instead of symbolic terms and using cryptographic algorithms instead of constructors and destructors. The first part of the proof entails the computational soundness of a process executing the abstraction $LMPC(l, t, F, sidc, adv, \widetilde{in})$. A computational implementation of the protocol, instead, should execute an actual MPC protocol.

In the second step of the proof, we show that for each l-function F, the computational execution of our abstraction $LMPC(l, t, F, sidc, adv, \widetilde{in})$ is indistinguishable from the execution of a ideal LMPC protocol I that solely comprises a single incorruptible machine.

The third step of the proof ensures that for a l-function $\widetilde{F}$ there is a protocol that securely realizes I in the UC framework, which ensures in particular that the uniformity properties of I carry over to the actual implementation.

These three steps allow us to conclude that for each abstraction $LMPC(l, t, F, sidc, adv, \widetilde{in})$ there exists an implementation such that the uniformity properties of any process P, carry over from the execution that merely executes a process P and executes $LMPC(l, t, F, sidc, adv, \widetilde{in})$ as a regular subprocess to the execution that communicates with upon each call of a subprocess $LMPC(l, t, F, sidc, adv, \widetilde{in})$. By leveraging the composability of the UC framework and the realization result for LMPC in the UC framework, we finally conclude that if a protocol based on our LMPC abstraction is robustly safe then there exists an implementation of that protocol that is computationally safe.

3.1   Computational Execution of a Process

We firstly give computational soundness definition of LSSS. Two properties are expected from a secret share proof in LSSS: knowledge of any t or more valid secret shares makes the secret easily computable (completeness), it is computationally infeasible to produce the secret with knowledge of any $t - 1$ or fewer valid secret shares(soundness):

Definition 5 *(Computational sound LSSS, Υ_{LSSS}). A symbolically-sound LSSS is a tuple of polynomial-time algorithms $(K, SCon, SCom, S)$ with the following properties (all probabilities are taken over the coin tosses of all algorithms and adversaries):*

1. *Completeness. Let a nonuniform polynomial-time adversary A be given. Let $(crs, simtd, extd) \leftarrow K(1^n)$. Let $(l, t, m) \leftarrow A(1^n, crs)$. Let $(ss_1, ss_2, ..., ss_l) \leftarrow SCon(l, t, m, crs)$, then with overwhelming probability in η, $SCom(l, t, ss_{i_1}, ss_{i_2}, ..., ss_{i_t}, crs) = m$.*
2. *Soundness. Let a nonuniform polynomial-time adversary A be given. Consider the following experiment parameterized by a bit c: Let $(crs, simtd, extd) \leftarrow K(1^n)$. Let $(l, t, m) \leftarrow A^{SCon(.^*),SCom(.^*)}(1^n, crs, simtd)$. Then let $(ss_1, ss_2, ..., ss_l) \leftarrow SCon(l, t, m, crs)$ if $c=0$ and $(ss_1, ss_2, ..., ss_l) \leftarrow S(l, t, simtd)$ if $c=1$. Let $guess = A^{SCon(.^*),SCom(.^*)}(crs, simtd, ss_{i_1}, ss_{i_2}, ..., ss_{i_r})$ with $r < t$. Let $P_c(\eta)$ denote the following probability: $P_c(\eta) := Pr[guess = m]$, then $|P_0(\eta) - P_1(\eta)|$ is negligible.*
3. *Length-regularity. Let polynomial-size circuit sequence $\widetilde{C}(.^*)$, and secret share parameters m_1, m_2 be given such that $|m_1| = |m_2|$. Let $(crs, simtd) \leftarrow K(1^n)$. Let $(ss_1) \leftarrow SCon(l, t, m_1, crs)$ and $(ss_2) \leftarrow SCon(l, t, m_2, crs)$. Then $|ss_1| = |ss_2|$ holds with probability 1.*

Since the applied pi-calculus only has semantics in the symbolic model (without probabilities and without the notion of a computational adversary), we need to introduce a notion of computational execution for symbolic protocol.

Our computational implementation of a symbolic protocol is a probabilistic polynomial-time algorithm that expects as input the symbolic protocol Π, a set of deterministic polynomial-time algorithms A for the constructors and destructors in Π, and a security parameter k. This algorithm executes the protocol by interacting with a computational adversary. In the operational semantics of the applied calculus, the reduction order is non-deterministic. This non-determinism is resolved by letting the adversary determine the order of the reduction steps. The computational execution sends the process to the adversary and expects a selection for the next reduction step.

Definition 6 *(Computational implementation of the symbolic model).* *We require that the computational implementation A of the symbolic model M has the following properties:*

1. *A is an implementation of M (in particular, all functions A_f ($f \in C \bigcup D$) are polynomial-time computable). For bitstring m, Type(m) denotes the type of m.*
2. *There are disjoint and efficiently recognizable sets of bitstrings representing the types nonces, and payload-strings. The set of all bitstrings of type nonce we denote $Nonces_k$. (Here and in the following, k denotes the security parameter.)*
3. *The functions $A_{SS_{l,t}}$ are length-regular. We call an function f length regular if $|m_i| = |m'_i|$ for $i = 1, 2, ..., n$ implies $|f(m_i)| = |f(m'_i)|$. All $m \in Nonces_k$ have the same length.*
4. *A_N for $N \in Nonce_{ag} \bigcup Nonce_{adv}$ returns a uniformly random $r \in Nonces_k$.*
5. *For all m, the image of $A_{SS_{l,t}}(m, r)$ is the sequence of the type $< secret\ share, Type(m) >$.*
6. *For all m, the image of $A_{COE_{l,t,i}}(A_{SS_{l,t}}(m, r))$ is of the type $< secret\ share, Type(m) >$.*
7. *For all $m_1, ..., m_t \in \{0,1\}^*$, if $m_1 = A_{COE_{l,t,i_1}}(A_{SS_{l,t}}(m, r))$, ..., $m_t = A_{COE_{l,t,i_t}}(A_{SS_{l,t}}(m, r))$, we have $A_{Combin_{l,t}}(m_1, ..., m_t) = m$. Else, $A_{Combin_{l,t}}(m_1, ..., m_t) = \perp$.*

Definition 7 *(Computational pi-calculus execution).* *Let Π be a closed process, A be a computational implementation of the symbolic model. Let Adv be an interactive machine called the adversary. We define the computational pi-calculus execution as an interactive machine $Exec_{\Pi,A}(1^k)$ that takes a security parameter k as argument and interacts with Adv:*

1. *Start. Let P be obtained from Π by deterministic α-renaming so that all bound variables and names in P are distinct. Let η and μ be a totally undefined partial functions from variables and names, respectively, to bitstrings. Let $a_1, a_2, ..., a_n$ denote the free names in P. For each i, pick $r_i \in Nonces_k$ at random. Set $\mu := \mu \bigcup \{a_1 := r_1, a_2 = r_2, ..., a_n := r_n\}$. Send $(r_1, r_2, ..., r_n)$ to Adv.*
2. *Main loop. Send P to the adversary and expect an evaluation context E from the adversary. Distinguish the following cases:*
   (a) $P = E[M(x) : P_1]$: Request two bitstrings c, m from the adversary. If $c = ceval_{\eta,\mu}(M)$, set $\eta := \eta \bigcup \{x := m\}$ and $P := E[P_1]$.

(b) $P = E[\nu a.P_1]$: Pick $r \in Nonces_k$ at random, set $P := E[P_1]$ and $\mu :=$ $\mu(a := r)$.

(c) $P = E[\widetilde{M}(N).P_1][M_2(x).P_2]$: If $ceval_{\eta,\mu}(M_1) = ceval_{\eta,\mu}(M_2)$, then set $P := E[P_1][P_2]$ and $\eta := \eta \bigcup \{x := ceval_{\eta,\mu}(N)\}$.

(d) $P = E[\text{let } x = D \text{ in } P_1 \text{ else } P_2]$: If $m := ceval_{\eta,\mu}(D)\perp$, set $\eta := \eta \bigcup$ $\{x := m\}$ and $P := E[P_1]$; Otherwise set $P := E[P_2]$.

(e) $P = E[!Q]$: Let Q' be obtained from Q by deterministic α-renaming so that all bound variables and names in Q' are fresh. Set $P := E[Q'|!Q]$. $P = E[\widetilde{M}(N).P_1]$: Request a bitstring c from the adversary. If $c = ceval_{\eta,\mu}(M)$, set $P := E[P_1]$ and send $ceval_{\eta,\mu}(N)$ to the adversary.

(f) In all other cases, do nothing.

For any interactive machine Adv, we define $Exec_{\Pi,A,Adv}(1^k)$ as the interaction between $Exec_{\Pi,A}(1^k)$ and Adv; the output of $Exec_{\Pi,A,Adv}(1^k)$ is the output of Adv.

In the preceding section, we have described the trace properties and the uniformity properties involving LSSS proofs. We firstly formulate our soundness result for trace properties, Namely, with overwhelming probability, a computational trace of computational pi-calculus execution is a computational instantiation of some symbolic Dolev-Yao trace.

We construct a interactive machine called simulator, which simulates against an adversary Adv the execution $Exec$ while actually interacting with $SExec$. The definition of such a simulator based on computational implementation A will be used for the definition for computational soundness for trace properties in the following.

Definition 8 *(Hybrid pi-calculus execution). The simulator Sim_A based on computational implementation A is constructed as follows: whenever it gets a term from the protocol, it constructs a corresponding bitstring and sends it to the adversary, and when receiving a bitstring from the adversary it parses it and sends the resulting term to the protocol.*

1. *Constructing bitstrings is done using a function β, parsing bitstrings to terms using a function τ. The simulator picks all random values and keys himself: For each protocol nonce N, he initially picks a bitstring r_N. He then translates, e.g., $\beta(N) := r_N$ and $\beta(SS_{l,t}(M,N)) := A_{SS_{l,t}}(r_M, r_N)$.*

2. *Translating back is also natural: Given $\widetilde{m} = \widetilde{r_N}$, we let $\tau(m_i) := COE_{l,t,i_1}$ ($SS_{l,t}(M,N)$), and if c is a LMPC result that can be decrypted as m using $A_{Com}(\widetilde{m})$, we set $\tau(c) := M$.*

Let Π be a closed process, Sim_A be a simulator based on computational implementation of the symbolic model A. Let Adv be an interactive machine called the adversary, we define the hybrid pi-calculus execution as an interactive machine $Exec_{\Pi,Sim_A}(1^k)$ that takes a security parameter k as argument and interacts with Adv. We also define $Exec_{\Pi,Sim_A,Adv}(1^k)$ as the interaction between $Exec_{\Pi,Sim_A}(1^k)$ and Adv; the output of $Exec_{\Pi,Sim_A,Adv}(1^k)$ is the output of Adv.

We stress that the simulator Sim does not have additional capabilities compared to a usual adversary against $Exec$. We then give the definition of the computational soundness for trace properties.

Definition 9 *(Computational soundness for trace properties). Let A be a computational implementation of the symbolic model and Sim_A be a simulator. If for every closed process Π, and polynomial-time interactive machine adv, A has to satisfy the following two properties:*

1. *Indistinguishability: $Exec_{\Pi,A,adv}(1^k) \approx Exec_{\Pi,Sim_A,adv}(1^k)$, which means the hybrid execution is computationally indistinguishable from the computational execution with any adversary.*
2. *Dolev-Yaoness: The simulator Sim_A never (except for negligible probability) sends terms t to the protocol with $S \mapsto t$ where S is the list of terms Sim_A received from the protocol so far.*

then A is a computationally sound model for trace properties.

Further, we give the definition of the computational soundness for uniformity properties. We rely on the notion of termination-insensitive computational indistinguishability (tic-indistinguishability) to capture that two protocols are indistinguishable in the computational world [22].

Definition 10 *(Tic-indistinguishability). Given two machines M, M' and a polynomial p, we write $Pr[(M \mid M') \Downarrow_{p(k)} x]$ for the probability that the interaction between M and M' terminates within $p(k)$ steps and M' outputs x. We call two machines A and B termination-insensitively computationally indistinguishable for a machine Adv ($A \approx_{tic}^{Adv} B$) if for for all polynomials p, there is a negligible function η such that for all z, a, b $\in [0,1]^*$ with $a \neq b$,*
$$Pr[(A(k) \mid Adv(k)) \Downarrow_{p(k)} a] + Pr[(B(k) \mid Adv(k)) \Downarrow_{p(k)} b] \preceq 1 + \eta(k)$$
Here, z represents an auxiliary string. Additionally, we call A and B termination-insensitively computationally indistinguishable $A \approx_{tic} B$ if we have $A \approx_{tic}^{Adv} B$ for all polynomial-time machines Adv.

Based on tic-indistinguishability, the definition of the computational soundness for uniformity properties is given as follows.

Definition 11 *(Computational soundness for uniformity properties). Let A be a computational implementation of the symbolic model and Sim_A be a simulator based on A. If A is a computational soundness model for trace properties, and for every uniform bi-process Π, $Exec_{left(\Pi),Sim_A}(1^k) \approx_{tic} Exec_{right(\Pi),Sim_A}(1^k)$, then A is a computational soundness model for uniformity properties.*

Lemma 1. *Assume that the PROG-KDM secure encryption scheme AE, the unforgeable signature scheme SIG, and the LSSS proof system Υ_{LSSS} satisfy the requirements in Definition 4. $(AE, SIG, \Upsilon_{LSSS})$ is a computationally sound model for uniformity properties.*

We now proceed to construct a generic ideal functionality $I_{sid,l,t,F}$ that serves as an abstraction of LMPC. This construction is parametric over the session identifier sid, and the function F to be computed. The ideal functionality receives the (secret) input message of party i from port $in^e_{i,sid}$ along with a session identifier. The input message is stored in the variable x_i and the state $state_i$ of i is set to input. Both the session identifier and the length of the received message are leaked to the adversary on port $out^a_{i,sid}$.

Definition 12 *(Ideal model for LMPC, I_{LMPC}). We construct an interactive polynomial-time machine $I_{sid,l,t,F}$, called the LMPC ideal functionality, which is parametric over a session identifier sid, LSSS threshold (l,t), and a poly-time algorithm F. Initially, the variables of $I_{sid,l,t,F}$ are instantiated as follows: $\forall i \in [1,n] : state_i := input, r_i := 1$. Upon an activation with message m on port p, $I_{sid,l,t,F}$, behaves as follows.*

1. *Upon $(in^e_{i,sid}(m, sid'))$ If $sid = sid'$ and $state_i = input$, then set $state_i := compute$ and $x_i := m$. If $state_i = input$, then send $(sid', |m|)$ on port $in^a_{i,sid}$.*
2. *Upon $(in^a_{i,sid}(deliver, sid'))$, if $\forall j \in [1;n] : state_j = compute$ and $r_i = r_j$, then compute $(y_1, y_2, ..., y_n) \leftarrow SCon(F(x_1, x_2, ..., x_n))$ and $\forall j \in [1,n] : state_j := deliver$. If $state_i = deliver$, set $r_i := r_i + 1$, $state_i := input$ and send y_i on port $out^e_{i,sid}$.*

For defining uniformity property-based computational soundness for ideal model $I_{sid,l,t,F}$, we modify the scheduling simulator Sim and hybrid computational execution as follow. Simulator $Sim_{A,I}$ simulates against an adversary Adv the execution $Exec$ while interacting with $SExec$, in which A is a computational implementation of the symbolic model and I is a family of LMPC ideal functionalities.

We also construct context, called $state_\gamma$. In our abstraction, every party of a LMPC can be in the following states: *init*, *input*, *compute*, and *deliver*. In the state *init*, the entire session is not initialized yet; in the state *input*, the party expects an input; in the state *compute*, the party is ready to start the main computation; and, in the state *deliver*, the party is ready to deliver secret share. These states are stored in a mapping state (which is maintained by $Sim_{A,I}$) such that $state_\gamma(i)$ is the state of party i in the session.

Definition 13 *(Hybrid pi-calculus execution with ideal model for LMPC). Let a mapping $state_\gamma(i)$ from internal session identifiers and party identifiers to states given. We assign a process to each state $state_\gamma$.*

1. *Upon receiving the initial process P, Enumerate every occurrence $LMPC(l, t, F, sidc, adv, \widetilde{in})$ with an internal session identifier γ, and tag this occurrence $LMPC(l, t, F, sidc, adv, \widetilde{in})$ in P with γ. Let initially $delivery(\gamma, i) := false$, and let $state_\gamma(i) := init$ for all $i \in [1, n]$. For any γ, let $corrupt(\gamma, i) := true$, if the corresponding in_i in $LMPC(l, t, F, sidc, adv, \widetilde{in})$ is free; otherwise let $corrupt(\gamma, i) := false$. In addition store all channel names in the partial mapping μ.*

2. *Main loop: Send P to the adversary Adv. Then, expect an evaluation context* E. *We distinguish the following cases for* E.

   (a) E *schedules the initialization and evaluation context is* $P = E[LMPC(l, t, F, sidc, adv, \widetilde{in})]$: *Set* $state_\gamma(i) := init$ *for all* $i \in [1, n]$. *The internal state of* $I_{session_{id}(\gamma),l,t,F}$ *is set to input.*

   (b) E *schedules an input to a corrupted party* i *and evaluation context is* $P = E[LMPC(l, t, F, sidc, adv, \widetilde{in})]$: *Request a bitstring* m *from the adversary. Check whether* $m = (c, s, input, m_0)$, $session_{id}(\gamma) := s$ *and* $state_\gamma(i) := input$, *If the execution accepts the channel name* c, *we proceed and check wether* $\mu(in_i)$ *is defined and* $c = \mu(in_i)$; *if* $\mu(in_i)$ *is not defined set* $\mu(in_i) := c$. *If the check fails, abort the entire simulation. Send* (m_0, s) *to* $I_{session_{id}(\gamma),l,t,F}$ *upon port* $in^e_{i,s}$, *Set* $state_\gamma(i) := compute$.

   (c) E *schedules an input to an honest party* i *and evaluation context is* $P = E[\widetilde{c}(x, s).Q][LMPC(l, t, F, sidc, adv, \widetilde{in})]$: *Check whether there is an* $i \in [1, n]$ *such that* $\mu(c) = \mu(in_i)$, $state_\gamma(i) := input$, *and* $session_{id}(\gamma) := \mu(s)$. *Send* $(\mu(x), \mu(s))$ *to* $I_{session_{id}(\gamma),l,t,F}$ *upon port* $in^e_{i,s}$, *Set* $state_\gamma(i) := compute$.

   (d) *Start the main computation upon the first delivery command for a party* i *and evaluation context:* $P = E[LMPC(l, t, F, sidc, adv, \widetilde{in})]$: *Check whether* $state_\gamma(i) := compute$ *for all* $i \in [1, n]$. *Set* $state_\gamma(i) := deliver$ *for all* $i \in [1, n]$.

   (e) *The delivery command for a party* i *is sent and evaluation context:* $P = E[LMPC(l, t, F, sidc, adv, \widetilde{in})]$: *Request a bitstring* $m = (c, s, deliver)$ *from the adversary. Check whether* $s = session_{id}(\gamma)$, *and* $state_\gamma(i) = deliver$ *and there is an* $i \in [1, n]$ *such that* $\mu(in_i) = c$. *Set* $delivery(\gamma, i) := true$, $state_\gamma(i) = input$, *receive* m', sid *from* $I_{session_{id}(\gamma),l,t,F}$ *on port* $out^e_{i,s}$ *and forward* m' *to adversary.*

   (f) *The output of party* i *is delivered to an honest party and evaluation context:* $P = E[c(x).Q][LMPC(l, t, F, sidc, adv, \widetilde{in})]$: *Check whether there is an* i *such that* $\mu(c) = \mu(in_i)$, *and* $state_\gamma(i) = deliver$. *Set* $delivery(\gamma, i) := true$, $state_\gamma(i) = input$, *receive* m', sid *from* $I_{session_{id}(\gamma),l,t,F}$ *on port* $out^e_{i,session_{id}(\gamma)}$ *and set* $\mu(x) = m'$.

Let Π be a closed process, $Sim_{A,I}$ be a simulator based on computational implementation of the symbolic model A and I be a family of Ideal model for LMPC. Let Adv be an polynomial-time interactive machine, we define the hybrid pi-calculus execution as an interactive machine $Exec_{\Pi,Sim_{A,I}}(1^k)$ that takes a security parameter k as argument and interacts with Adv. We also define $Exec_{\Pi,Sim_{A,I},Adv}(1^k)$ as the interaction between $Exec_{\Pi,Sim_{A,I}}(1^k)$ and Adv; the output of $Exec_{\Pi,Sim_{A,I},Adv}(1^k)$ is the output of Adv.

Then we can give the definition of uniformity property-based computational soundness for LMPC ideal model.

Definition 14 *(Uniformity property-based computational soundness for LMPC ideal model,* I_{LMPC}*). Let* A *be computational implementation of the symbolic* *model and I be a family of ideal model for LMPC. If for every closed process Π, and polynomial-time interactive machine adv, $Exec_{\Pi, Sim_{A, I_{LMPC}}, adv}(1^k) \approx Exec_{\Pi, Sim_A, adv}(1^k)$, then (A, I) is a computationally sound ideal model for LMPC.*

We can get the computation soundness result for ideal model for LMPC.

Lemma 2. *Assume that the PROG-KDM secure encryption scheme AE, the unforgeable signature scheme SIG, and the LSSS proof system Υ_{LSSS} satisfy the requirements in Definition 5, the family of ideal model for LMPC I_{LMPC} satisfy the requirement in Definition 12, (AE, SIG, Υ_{LSSS}, I_{LMPC}) is a computationally soundness ideal model for LMPC.*

3.2 Computational Soundness Results

We now state the main computational soundness result of this work: the robust safety of a process (specifically uniformity properties) using non-interactive primitives and our LMPC abstraction carries over to the computational setting, as long as the non-interactive primitives are computationally sound. This result ensures that the verification technique from Sect. 3 provides computational safety guarantees. We stress that the non-interactive primitives can be used both within the LMPC abstractions and within the surrounding protocol.

In order to realize the definition of computation soundness for uniformity properties in UC framework, we firstly give a strong definition for the security requirement of UC realization.

Definition 15 *(Computational soundness of UC realization for LMPC). Let A be a computational implementation of the symbolic model, and I be an ideal functionality family, we say that real protocol family ρ securely realizes I if*

1. *(Computational soundness for trace properties) For any adversary Adv, there exists an ideal-process adversary S such that for any LMPC Π and environment Z,*
   $Exec_{I(\Pi), S, Z} \approx Exec_{\rho(\Pi), Adv, Z}$.
2. *(Computational soundness for uniformity properties) For any LMPC Π_1, Π_2 and environment Z, if for any ideal-adversary S such that*
   $Exec_{I(\Pi_1), S, Z} \approx Exec_{I(\Pi_2), S, Z}$,
   then for any adversary Adv,
   $Exec_{\rho(\Pi_1), Adv, Z} \approx Exec_{\rho(\Pi_2), Adv, Z}$

Finally, we can get the computation soundness result for our abstraction for LMPC.

Theorem 1. *Assume that the PROG-KDM secure encryption scheme AE, the unforgeable signature scheme SIG, and the LSSS proof system LSSS satisfy the security requirements in Definition 5, also the family of ideal model for LMPC I_{LMPC} satisfy the requirement in Definition 12, Assume the existence of sub-exponentially secure one-way functions. Then for all ideal function $I \in \mathcal{I}_{LMPC}$, there exists a non-trivial real LMPC protocol $\rho \in \rho_{LMPC}$ secure realizes I in the presence of malicious, static adversaries.*

4   Conclusions

We have presented the first computational soundness theorem for multi-party computation based on linear secret-sharing scheme (LMPC). This allows to analyze protocols in a simple symbolic model supporting encryptions, signatures, and secret-sharing schemes; the computational soundness theorem then guarantees that the uniformity properties shown in the symbolic model carry over to the computational implementation.

References

1. Shamir, A.: How to share a secret. Commun. ACM **22**(11), 612–613 (1979)
2. Feldman, P.: A practical scheme for non-interactive verifiable secret sharing. In: Proceedings of the 28th Annual IEEE Symposium on Foundations of Computer Science, pp. 427–437 (1987)
3. Shoup, V.: Practical threshold signatures. In: Preneel, B. (ed.) EUROCRYPT 2000. LNCS, vol. 1807, pp. 207–220. Springer, Heidelberg (2000)
4. He, A.J., Dawson, E.: Multistage secret sharing based on one-way function. Electron. Lett. **30**(9), 1591–1592 (1994)
5. Chien, H.-Y., Tseng, J.K.: A practical (t, n) multi-secret sharing scheme. IEICE Trans. Fundam. Electron. Commun. Comput. **83–A**(12), 2762–2765 (2000)
6. Shao, J., Cao, Z.F.: A new efficient (t, n) verifiable multi-secret sharing (VMSS) based on YCH scheme. Appl. Math. Comput. **168**(1), 135–140 (2005)
7. Zhao, J., Zhang, J., Zhao, R.: A practical verifiable multi-secret sharing scheme. Comput. Stand. Interfaces **29**(1), 138–141 (2007)
8. Yang, C.C., Chang, T.Y., Hwang, M.S.: A (t, n) multi-secret sharing scheme. Appl. Math. Comput. **151**, 483–490 (2004)
9. Cramer, R., Damgård, I.B., Maurer, U.M.: General secure multi-party computation from any linear secret-sharing scheme. In: Preneel, B. (ed.) EUROCRYPT 2000. LNCS, vol. 1807, pp. 316–334. Springer, Heidelberg (2000)
10. Abadi, M., Baudet, M., Warinschi, B.: Guessing attacks and the computational soundness of static equivalence. In: Aceto, L., Ingólfsdóttir, A. (eds.) FOSSACS 2006. LNCS, vol. 3921, pp. 398–412. Springer, Heidelberg (2006)
11. Abadi, M., Fournet, C.: Mobile values, new names, and secure communication. In: Proceedings of the 28th Symposium on Principles of Programming Languages (POPL), pp. 104–115. ACM Press (2001)
12. Abadi, M., Rogaway, P.: Reconciling two views of cryptography (the computational soundness of formal encryption). J. Crypt. **15**(2), 103–127 (2002)
13. Backes, M., Maffei, M., Mohammadi, E.: Computationally sound abstraction and verification of secure multi-party computations. In: Proceedings of ARCS Annual Conference on Foundations of Software Technology and Theoretical Computer Science (FSTTCS) (2010)
14. Backes, M., Hofheinz, D., Unruh, D.: A general framework for computational soundness proofs or the computational soundness of the applied pi-calculus. IACR ePrint Archive 2009/080 (2009)
15. Backes, M., Bendun, F., Unruh, D.: Computational soundness of symbolic zero-knowledge proofs: weaker assumptions and mechanized verification. In: Basin, D., Mitchell, J.C. (eds.) POST 2013 (ETAPS 2013). LNCS, vol. 7796, pp. 206–225. Springer, Heidelberg (2013)

16. Backes, M., Malik, A., Unruh, D.: Computational soundness without protocol restrictions. In: CCS, pp. 699–711. ACM Press (2012)
17. Kusters, R., Tuengerthal, M.: Computational soundness for key exchange protocols with symmetric encryption. In: Proceedings of the 16th ACM Conference on Computer and Communications Security (CCS), pp. 91–100. ACM Press (2009)
18. Canetti, R., Lindell, Y., Ostrovsky, R., Sahai, A.: Universally composable two-party and multiparty secure computation. In: Proceedings of the 34th Annual ACM Symposium on Theory of Computing (STOC), pp. 494–503. ACM Press (2002)
19. Comon-Lundh, H., Cortier, V.: Computational soundness of observational equivalence. In: Proceedings of the 16th ACM Conference on Computer and Communications Security (CCS), pp. 109–118. ACM Press (2008)
20. Comon-Lundh, H., Cortier, V., Scerri, G.: Security proof with dishonest keys. In: Degano, P., Guttman, J.D. (eds.) Principles of Security and Trust. LNCS, vol. 7215, pp. 149–168. Springer, Heidelberg (2012)
21. Canetti, R.: Herzog: universally composable symbolic security analysis. J. Cryptol. **24**(1), 83–147 (2011)
22. Backes, M., Mohammadi, E., Ruffing, T.: Computational soundness results for ProVerif. bridging the gap from trace properties to uniformity. In: Kremer, S., Abadi, M. (eds.) POST 2014 (ETAPS 2014). LNCS, vol. 8414, pp. 42–62. Springer, Heidelberg (2014)
23. Canetti, R., Feige, U., Goldreich, O., Naor, M.: Adaptively secure multi-party computation. In: Proceedings of 28th STOC, pp. 639–648 (1996)
24. Canetti, R., Fischlin, M.: Universally composable commitments. In: Kilian, J. (ed.) CRYPTO 2001. LNCS, vol. 2139, pp. 19–40. Springer, Heidelberg (2001)
25. Canetti, R., Rabin, T.: Universal composition with joint state. Cryptology ePrint Archive. Report 2002/047 (2002). http://eprint.iacr.org/

Attribute-Based Signatures
with Controllable Linkability

Miguel Urquidi[1]([⊠]), Dalia Khader[2], Jean Lancrenon[1], and Liqun Chen[3]

[1] Interdisciplinary Centre for Security, Reliability and Trust (SnT),
Luxembourg, Luxembourg
{miguel.urquidi,jean.lancrenon}@uni.lu
[2] POST Telecom PSF, Mamer, Luxembourg
dalia.khader@post.lu
[3] Hewlett Packard Labs, Bristol, UK
liqun.chen@hpe.com

Abstract. We introduce Attribute-Based Signatures with Controllable Linkability ABS-CL. In general, Attribute-Based Signatures allow a signer who possesses enough attributes to satisfy a predicate to sign a message without revealing either the attributes utilized for signing or the identity of the signer. These signatures are an alternative to Identity-Based Signatures for more fine-grained policies or enhanced privacy. On the other hand, the Controllable Linkability notion introduced by Hwang et al. [14] allows an entity in possession of the linking key to determine if two signatures were created by the same signer without breaking anonymity. This functionality is useful in applications where a lower level of anonymity to enable linkability is acceptable, such as some cases of vehicular ad-hoc networks, data mining, and voting schemes. The ABS-CL scheme we present allows a signer with enough attributes satisfying a predicate to sign a message, while an entity with the linking key may test if two such signatures were created by the same signer, all without revealing the satisfying attributes or the identity of the signer.

Keywords: Anonymity · Privacy · Group signatures · Attribute-Based Signatures · Linkability · Controllable Linkability

1 Introduction

In traditional digital signature schemes, the recipient of a signature is convinced that a particular signer associated with some identity has authenticated the received message. Attribute-Based Signatures (ABS) were first proposed by Maji et al. [17], in which messages are signed with respect to a signing policy expressed as a predicate. In an ABS scheme a valid signature can be generated only if the signer possesses enough attributes to satisfy the predicate, and the signature does not reveal the identity of the signer nor the attributes she used to create it [17,18]. Hence, the recipient is instead convinced that some signer possessing

© Springer International Publishing Switzerland 2016
M. Yung et al. (Eds.): INTRUST 2015, LNCS 9565, pp. 114–129, 2016.
DOI: 10.1007/978-3-319-31550-8_8

enough attributes to satisfy the predicate has authenticated the message, as opposed to a particular identity.

As explained in [2,13], privacy is characterized by the notions of anonymity and unlinkability. The former refers to the property that given a single signature the identity of the signer is concealed in the message, while the latter refers to the property that given two signatures an unauthorized entity cannot determine if they were created by the same signer.

Consider a situation in which users are permitted to join a multi-party computation if and only if the possess certain attributes. An instance of this could be a voting scheme where participants satisfy a predicate over their attributes. Attribute-Based Signatures are a natural candidate to attempt to solve this problem, but participants could easily abuse the system by double-voting, or by signing messages on behalf of other users (who might not have enough attributes to satisfy the predicate themselves). Since there is no relation amongst signatures one cannot prevent or detect the previously mentioned cases. Therefore, we would like to introduce a notion of linkability to attribute-based signatures. We want to add such a property while preserving a sense of anonymity, hence we will be adjusting the security notions corresponding to our functionality.

Intuitively, given two signatures $\sigma_1 \leftarrow \text{Sign}(m_1, SK_1)$ and $\sigma_2 \leftarrow \text{Sign}(m_2, SK_2)$, on messages m_1 and m_2, and using signing keys SK_1 and SK_2, we want a functionality $\text{Link}(\sigma_1, \sigma_2)$ that outputs 1 if $SK_1 = SK_2$, and 0 otherwise, i.e., we want to indicate if the signatures were created by the same signer, while preserving anonymity.

In essence the problem that we want to focus on is that of designing an ABS scheme with a linkability functionality. However - in contrast with the functionality considered in [7] - we want linkability to be available only for certain trusted entities.

1.1   Related Work

There have been several variants of ABS proposed in the literature, including schemes supporting different expressiveness for the predicates, such as non-monotonic [19], monotonic [18], and threshold predicates [5,9,12,22]. The case of multiple attribute authorities has been considered [7,17,19], where each attribute authority would be responsible for a subset of the universe of possible attributes. The multi-authority case first relied on the existence of a central trusted authority, and soon decentralized schemes were proposed [5,8,20].

Various concerns with respect to ABS have been addressed in the literature, such as revocation [16,23]. The notion of traceability has been covered [8,10], which adds a mechanism allowing a tracing authority to recover the identity of the signer if needed. A recent work on linkability of ABS [7] allows the signer to decide at signing time if her signatures can be linked with respect to a given recipient, therefore referred to as user-controlled linkability.

Aside from ABS, there are two main cryptographic solutions to preserve signer privacy in the literature. These are pseudonym systems [21] and group signatures [6]. The approach of [21] supports some anonymity, but signature links are publicly verifiable by anyone. Hence, this may be thought of as a *public* linkability functionality. In [6], the capability to link is given only to a trusted entity, allowing more control over link disclosure. However, involving a single entity to check links introduces a bottleneck and requires on-line processing. This may be considered a *private* linkability functionality.

We will focus on a system where information linking the signatures is not publicly available. Instead, it is controlled by trapdoors, and the linking functionality may be delegated to reduce the bottleneck effect. We intend to add this functionality to an ABS scheme.

1.2   Contribution

Our contribution is a formal security model for Attribute-Based Signatures with Controllable Linkability (ABS-CL), where signatures can be anonymously linked if in possession of a linking key. This key is managed by a trusted entity called a *linker*.

One ABS scheme with a linkability functionality has already been proposed in the literature [7]. It does not require a linking key. Instead, anyone can determine if two signatures were created by the same signer with respect to a certain recipient tag. In that sense, the scheme in [7] is publicly verifiable, and the linking functionality is comparable to that of Direct Anonymous Attestation schemes [4]. Our contribution is the first Controllable Linkability scheme in the ABS setting, where a *linking key is needed* to determine if two signatures were created by the same signer. Hence, our scheme is rather privately verifiable, and the added functionality is comparable to that in [13,14].

Depending on the application environment, the privacy needs of users and service providers may vary. For example, in ABS [17,18] neither identifying information nor linking information is revealed from the signatures, while in [7] no identifying information is revealed, but there is a user-provider negotiation on whether there will be no information linking signatures or if linkability may be checked by anyone. Our scheme resides conceptually in-between these two: the linking information is revealed not to everyone, but only to those entities possessing the linking-key.

Thus, our scheme has applications in existing services where ABS could be used and where linkability is required, such as vehicular ad-hoc networks, and data mining, to mention some. In these application scenarios, it is unnecessary, and even undesirable for privacy reasons, that linkability be publicly possible. Instead, users may authenticate through ABS, hiding their identifiable information, but still revealing links between signatures to the appropriate entities (authorities or data mining service providers) authorized to hold linking keys.

It may also be implemented in new services. One could imagine for example, as mentioned previously, a voting scheme where voters are required to possess certain attributes to participate, but where some authorized entities are in charge of preventing double-voting. This clearly cannot be achieved by ABS alone, since linking is impossible. Nor is it achievable by [7], since such signatures either reveal no linking information or reveal it to everyone, including potential coercers.

Our ABS-CL scheme and security notions are described in the case of multiple attribute authorities. Finally, we provide a provably secure instantiation of the scheme, with a single attribute authority. The construction is based on one of the ABS instantiations in [18], to which we add the linkability functionality with techniques inspired by [14,15].

1.3   Organization

The remainder of this paper is organized as follows. In Sect. 2 we establish some preliminaries and computational assumptions. In Sect. 3, we define the concept of Attribute-Based Signatures with Controllable Linkability, together with its syntax, and security notions. We propose a construction of an ABS-CL scheme in Sect. 4, followed by details of its correctness and sketches for the security proofs in Sect. 5. We conclude the paper in Sect. 6.

2   Preliminaries

We introduce briefly the concepts needed for our scheme.

Bilinear Pairings. Groups with Bilinear Pairings.

Let $\mathbb{G}, \mathbb{H}, \mathbb{G}_T$ be cyclic multiplicative groups of order p prime. Let g and h be generators of $\mathbb{G}$ and $\mathbb{H}$, respectively.

Then, $e : \mathbb{G} \times \mathbb{H} \to \mathbb{G}_T$ is a bilinear pairing if $e(g,h)$ is a generator of $\mathbb{G}_T$, and $e(g^x, h^y) = e(g,h)^{xy}$, for all x, y.

NIZK Proofs. Non-Interactive Zero-Knowledge Proofs.

A Non-Interactive Zero-Knowledge scheme Π is comprised of the following main algorithms:

Π.Setup: Outputs a common reference string CRS and extraction key xk.

Π.Prove: On input (CRS, Φ, w), where Φ is a Boolean formula and $\Phi(w) = 1$, it outputs a proof π.

Π.Verify: On input (CRS, Φ, π), it outputs a Boolean value.

Π.Extract: On input (CRS, xk, π) it outputs a witness w.

Π.SimSet: Outputs a common reference string CRS_{sim} and a trapdoor θ.

Π.SimProve: On input $(CRS_{sim}, \Phi, \theta)$, for a possibly false statement Φ, it outputs a simulated proof π_{sim} but without any witness.

In addition, we require completeness, soundness, and zero-knowledge. Completeness requires that honestly generated proofs be accepted, while soundness that it is infeasible to produce convincing proofs for false statements, and zero-knowledge that proofs reveal no information about the witness used.

2.1   Computational Assumptions

For the following definitions, let $\mathbb{G}, \mathbb{H}, \mathbb{G}_T$ be cyclic multiplicative groups of order p prime; let g, and h, be generators of $\mathbb{G}$ and $\mathbb{H}$, respectively; and let $e : \mathbb{G} \times \mathbb{H} \to \mathbb{G}_T$ be a bilinear pairing. Hereby are the assumptions needed for the security of our proposed protocol.

Definition 1. *DDH. Decisional Diffie-Hellman*

Let x and y, be selected randomly from $\mathbb{Z}_p^*$. The Decisional Diffie-Hellman (DDH) problem is to decide if $z = xy$, for a given triplet $(g^x, g^y, g^z) \in \mathbb{G}^3$.

We say that the DDH assumption holds in $\mathbb{G}$ if any polynomial-time adversary A has a negligible advantage in solving the DDH problem.

Definition 2. *q-SDH. q-Strong Diffie-Hellman*

Let x be selected randomly from $\mathbb{Z}_p^*$. The q-Strong Diffie-Hellman (q-SDH) problem is to compute a pair $(g^{1/(x+y)}, y)$, for a given (q+3)-tuple $(g, g^x, \ldots, g^{x^q}, h, h^x)$.

We say that the q-SDH assumption holds in $(\mathbb{G}, \mathbb{H})$ if any polynomial-time adversary A has a negligible advantage in solving the q-SDH problem.

Definition 3. *SXDH. Symmetric eXternal Diffie-Hellman*

Let $\mathbb{G}, \mathbb{H}, \mathbb{G}_T$ be as described above, we say that the Symmetric eXternal Diffie-Hellman (SXDH) assumption holds in $(\mathbb{G}, \mathbb{H})$ if the Decisional Diffie-Hellman assumption holds in both groups $\mathbb{G}$ and $\mathbb{H}$.

3   ABS with Controlled-Linkability

An Attribute-Based Signature with Controllable Linkability scheme is parametrized by a universe of attributes $\mathbb{A}$, and message space $\mathbb{M}$. Users are described by *attributes* rather than their identity. The scheme allows a user, whose attributes satisfy a *predicate* of his choice, to create a valid signature of a message under that predicate. Attribute Authorities are required to distribute attribute signing keys appropriately. The predicate is up to the choice of the signer. The identity of the signer should remain hidden amongst all those that may possess enough attributes to satisfy the predicate.

Let $\mathbb{A}$ be the universe of possible attributes, and let $\mathbb{M}$ be the message space. An attribute predicate ψ over $\mathbb{A}$ is a monotone Boolean function, whose inputs are attributes over $\mathbb{A}$. We say that an attribute set $\mathcal{A} \subset \mathbb{A}$ satisfies the policy ψ if $\psi(\mathcal{A}) = 1$. Also, we say that an attribute $a \in \mathcal{A}$ is needed to satisfy the predicate ψ if $\psi(\mathbb{A} \backslash \{a\}) = 0$.

We will associate each user with a tuple (i, τ_i), containing her index i, and a unique tag τ_i. Similarly, we associate each attribute authority with a tuple (j, APK_j, ASK_j), containing his index j, public key APK_j, and secret key ASK_j. We will use PK to denote the global public key $PK = (TPK, \bigcup APK_j)$. We will often assume the security parameter $\mathcal{K}$ to be implicit in the algorithms, and thus omit it as an argument.

The scheme assumes that there are several authorities with various levels of trust: An *issuer*, which sets up the whole system and must be fully trusted; a *linker*, given the ability to link signatures; and potentially multiple *attribute authorities*, some of which may be corrupted. In our scheme, we focus on the case of only one attribute authority.

The scheme consists of the following algorithms:

- **TSetup**$(\mathcal{K})$. Outputs public parameters TPK, and a master linking key mlk, which is delegated to the linking authority. This algorithm is run by the issuer, and serves to set up the whole system.
- **AttSet**$(\mathcal{K})$. Outputs an attribute authority public key/private key pair (APK, ASK). This is run by the attribute authority. APK is made public, and ASK will be used to generate user signing keys from attributes managed by this authority.
- **AttGen**$(ASK, \mathcal{A}, \tau)$. Outputs a user signing key $SK_{\mathcal{A},\tau}$, corresponding to set $\mathcal{A}$ and a string uniquely associated to the user requesting the signing key. This algorithm is run by the attribute authority to generate attribute-associated signing keys for users. τ is central to our linking function.
- **AttVer**$(APK, \mathcal{A}, \tau, SK_{\mathcal{A},\tau})$. Outputs a Boolean value. This algorithm is needed to verify that keys provided by attribute authorities to users are valid with respect to the attribute authorities' public keys.
- **Sign**$(PK, \psi, m, SK_{\mathcal{A},\tau})$. Where $\psi(\mathcal{A}) = 1$, outputs a signature σ of message m. This is the signing algorithm itself, run by a user with a signing key that is adequate relative to the predicate.
- **Veri**(PK, ψ, m, σ). Outputs a Boolean value. This is just signature verification.
- **Link**$(PK, (m_1, \psi_1, \sigma_1), (m_2, \psi_2, \sigma_2), mlk)$. If σ_1 and σ_2 are valid signatures of messages m_1 and m_2 under predicate ψ, it outputs a Boolean value. This is the linking algorithm. If the signatures are not valid, linking is a moot point. If they are both valid, this algorithm should indicate whether the same user created them or not. Note that the linking key mlk is required as input.

The correctness of the ABS-CL scheme and the property of linkability, are formally defined next.

Definition 4. *Correctness (of signing)*

Correctness (of signatures) requires that signatures generated by an honest user should be verified correctly. That is, for all $TPK \leftarrow TSetup$, all public keys $APK \leftarrow AttSet$, all messages $m \in \mathbb{M}$, all policies ψ, all attribute sets $\mathcal{A} \subset \mathbb{A}$ such that $\psi(\mathcal{A}) = 1$, all signing keys $SK_{\mathcal{A},\tau} \leftarrow AttGen(ASK, \mathcal{A}, \tau)$, and all signatures $\sigma \leftarrow Sign(PK, \psi, m, SK_{\mathcal{A},\tau})$, we have that $Veri(PK, \psi, m, \sigma) = 1$.

Definition 5. *Correctness (of linking)*

Correctness (of linkability) requires that signatures produced by the same user should link correctly. That is, for valid signatures σ_i produced by τ_i, and σ_j produced by τ_j, where $\tau_i = \tau_j$, we have that

$$Link(PK, (m_i, \psi_i, \sigma_i), (m_j, \psi_j, \sigma_j), mlk) = 1.$$

3.1 Unforgeability

Unforgeability refers to the property that a valid signature σ for a message m under a policy ψ cannot be efficiently created from public data alone. Only users that possess a signing key $SK_{\mathcal{A},\tau}$ for an attribute set $\mathcal{A}$ that satisfies $\psi(\mathcal{A}) = 1$ and tag τ should be able to create valid signatures.

We want to prevent users from signing under predicates ψ, which they do not hold enough attributes to satisfy even in cases of user collusions. These are called *signature forgeries*.

In addition, we want to prevent an adversary from producing a valid signature that was not created using the secret keys from an honest user, but does link to that honest user. We call this a *link forgery*.

Definition 6. *Unforgeability*

An ABS-CL scheme is said to be unforgeable if the advantage of a polynomial-time adversary A is negligible in the following experiment:

Setup:

The challenger runs $(TPK, mlk) \leftarrow TSetup(\mathcal{K})$, and gives (TPK, mlk) to the adversary.

Capabilities of the Adversary:

1. *Can ask for the signing keys of any attribute set under any attribute authority or user identity of his choice.*
2. *Can ask for the secret keys of any attribute authority of his choice.*
3. *Has access to a signing oracle that he may query on behalf of any user, on messages and predicates of his choice.*

Output:

Case "forging signature": He outputs a signature σ^* on message m^* with respect to predicate ψ^*. With the restrictions that (m^*, ψ^*) was never queried to the signing oracle, and there is at least one non-corrupted attribute needed to satisfy ψ^*, i.e., there exists $a^* \notin CAttr$ such that $\psi^*(\mathbb{A} \backslash \{a^*\}) = 0$.

He wins the game if σ^* is a valid signature.

Case "forging link": With the same restrictions as the previous game, he outputs a signature σ^* on message m^* with respect to predicate ψ^*, along with a message m', a predicate ψ', and the τ of an honest user.

He wins the game if

$$Link(PK, (m^*, \psi^*, \sigma^*), (m', \psi', \sigma'), mlk) = 1,$$

where σ' is a valid signature of message m' w.r.t. ψ', produced by the user associated to τ.

3.2 Anonymity

Anonymity refers to the property that an honest user τ who created a signature σ, cannot be efficiently identified as the author of the signature. One can only infer some information about the identity through the linking functionality, when one possesses the master linking key mlk and a signature by user τ.

This notion requires that a signature should reveal no information about the identity of the signer, nor the attributes used to sign. Given a valid signature σ, generated by one of two honest users chosen by the adversary, it should be a hard problem to correctly guess the generator of σ.

For clarity we will be splitting this notion into two games, depending on whether or not the adversary has access to the master linking key. When the adversary does have mlk, he is given a signature from one of two possible honest users, and we want to ensure that he is neither able to gain any information regarding the attributes used, nor is he able to distinguish which of the two signers created the signature, *provided he has not previously obtained signatures from said signers*. When the adversary does not have mlk, we want the same guarantees, this time *regardless of any signatures he may have already observed*.

We consider first the case when the adversary does not have access to mlk. If the adversary were able to gain information on the attributes used in a signature, then by careful choice of $\mathcal{A}_0$ and $\mathcal{A}_1$ on the challenge, he would be able to distinguish between the two signatures and gain some advantage in the game.

Definition 7. *Anonymity (without the master linking key)*
An ABS-CL scheme is said to be anonymous if the advantage of a polynomial-time adversary A is negligible in the following experiment:

Setup:
The challenger runs $(TPK, mlk) \leftarrow TSetup(\mathcal{K})$, and gives TPK to the adversary.

Capabilities of the Adversary:

1. *The adversary has full control over all attribute authorities, and has access to their secret keys.*
2. *Has access to a linking oracle, but may not query it on the issued challenge.*

Challenge:
The adversary may use his capabilities, and outputs

$$(\psi, m, \tau_0, \mathcal{A}_0, SK_{\mathcal{A}, \tau_0}, \tau_1, \mathcal{A}_1, SK_{\mathcal{A}, \tau_1}, APK),$$

such that $AttVer(APK, \mathcal{A}, \tau_i, SK_{\mathcal{A}, \tau_i}) = 1$, and $\psi(\mathcal{A}_i) = 1$, for $i = 0, 1$.
The challenger chooses randomly $b \leftarrow \{0, 1\}$, and the adversary receives a valid signature σ_b produced from $(\mathcal{A}_b, SK_{\mathcal{A}_b, \tau_b})$.
The adversary may then continue using his capabilities.

Output:
Finally, the adversary outputs b^*, and wins the game if $b^* = b$.

Now, we cover the security notion that should be achieved when the adversary possesses the master linking key mlk. This corresponds to the case of an honest-but-curious linker, where we assume that he is not colluding with any attribute authority, and we want to achieve that from signatures of non-revealed users the linker is not able to infer any information on the identity of the signer.

Definition 8. *Anonymity (with the master linking key)*

An ABS-CL scheme is said to be anonymous if the advantage of a polynomial-time adversary A *is negligible in the following experiment:*

Setup:

The challenger runs $(TPK, mlk) \leftarrow TSetup(\mathcal{K})$, *and gives* (TPK, mlk) *to the adversary.*

Capabilities of the Adversary:

1. Has access to a signing oracle that he may query in behalf of any user, on messages and predicates of his choice.

Challenge:

The adversary may use his capabilities, and outputs $(\psi, m, \tau_0, \mathcal{A}_0, \tau_1, \mathcal{A}_1)$, *where* $\psi(\mathcal{A}_i) = 1$ *for* $i = 0, 1$.

The restrictions are that τ_0 *and* τ_1 *must be honest users, and neither* τ_0 *nor* τ_1 *were queried to the signing oracle.*

The challenger chooses randomly $b \leftarrow \{0, 1\}$, *and the adversary receives a valid signature* σ_b *produced from* $(\tau_b, \mathcal{A}_b)$.

The adversary may then continue using his capabilities, while respecting the indicated restrictions.

Output:

The adversary outputs b^*, *and wins the game if* $b^* = b$.

4   Construction of ABS-CL

Intuition. Our construction is basically a modification of the first scheme from [18], in order to allow it to support the kind of linkability feature we want. Thus, the intuition behind our scheme is quite similar to theirs. We recap it now, in the case of a single attribute authority.

We first look at how the knowledge of attributes is conveyed through the signing process. Letting a be an attribute and τ be the tag associated to a unique user, an attribute key is a simple *digital signature* on the message (a, τ) under the attribute authority's private key, which is a signing key. This binds the attribute a to the user via the user-specific tag τ. In order to create a signature under some predicate ψ, the user basically proves in a witness indistinguishable way that it knows attribute keys corresponding to a subset of attributes sufficient to satisfy ψ. In particular, the proof must convince the verifier that the attribute keys are valid signatures, all binding necessary attributes to *the same* τ, in order to prevent collusions of users. Therefore, a signature is basically a NIZK proof.

We now explain how to bind a message m to the proof. This is done by slightly modifying the predicate ψ to satisfy. Instead of proving that she knows enough of the necessary signatures on attributes bound to τ in order to satisfy ψ, the signer proves that she knows *either* those, *or* a signature under *the trustee's signing key* on a unique string - referred to in the sequel as a *pseudo-attribute* - that encodes the pair (m, ψ). Of course, signatures of the latter form are assumed to *never be issued by the trustee*. Therefore, any verifier is convinced that the signer indeed possesses an adequate set of attribute keys, satisfying ψ. What has been gained however, is that the proof will only be valid when *verified with message m*, since m is now encoded in the modified predicate.

We now finally give some intuition as to how the linking mechanism works. The idea is to append to the proof an encryption of a piece of data $D(\tau)$ that uniquely determines τ. The NIZK proof is also modified to show that the signer created this encryption, and that the τ value involved is the same as that in the attribute keys. This encryption is instantiated in such a way that anybody in possession of the linking key can verify whether two ciphertexts contain the same value of $D(\tau)$, without actually obtaining the value itself. However, anybody not possessing this key still sees nothing. This preserves the anonymity of the signature fully, when not holding the linking key. Our instantiation accomplishes this using bilinear maps, using a technique inspired by [14,15]. Essentially, the value $D(\tau)$ will for us simply be C^τ for a group element C.

Of course, as has already been mentioned, we have to accept that some anonymity is sacrificed to the linking entity. It is inevitable that any form of linking functionality will do this.

Construction. We denote the universe of attributes with $\mathbb{A}$, and the universe of pseudo-attributes with $\mathbb{A}'$. In our construction we assume that $\mathbb{A} \cup \mathbb{A}' \subset \mathbb{Z}_p^*$ and that $\mathbb{A} \cap \mathbb{A}' = \emptyset$.

We now proceed to explain our instantiation of an ABS-CL scheme. This construction is for the case of one honest attribute authority.

The signature scheme we use both at the trustee and to run the attribute key generation is the Boneh-Boyen scheme [3]. It is known to be secure under the q-SDH assumption in bilinear groups. As for the NIZK, we can use that of Groth-Sahai [11], which can be instantiated under the SXDH assumption. Using this instantiation makes sense, as it will be apparent that we need DDH to hold in the group $\mathbb{G}$ in a bilinear system $(g, h, \mathbb{G}, \mathbb{H}, \mathbb{G}_T, e)$.

- **TSetup($\mathcal{K}$).** Chooses the bilinear map parameters $(g, h, \mathbb{G}, \mathbb{H}, \mathbb{G}_T, e)$, computes a common reference string CRS for the Groth-Sahai proof system, and chooses a random $\xi \leftarrow \mathbb{Z}_p^*$ and a random $\lambda \leftarrow \mathbb{G}^*$. Then, computes $\Lambda := \lambda^\xi$ and the linking key $mlk := h^\xi$. Finally, runs the Boneh-Boyen signature key generation to obtain (TVK, TSK).

  Outputs $TPK = (g, h, \mathbb{G}, \mathbb{H}, \mathbb{G}_T, e, CRS, \lambda, \Lambda, TVK)$, mlk, and TSK. TPK is public, mlk is to be given to a linking authority, and TSK remains secret.
- **AttSet($\mathcal{K}$).** Chooses a random $ASK := (b, c, d) \leftarrow \mathbb{Z}_p^{*3}$ and computes $APK := (B, C, D) := (g^b, g^c, g^d) \in \mathbb{G}^3$.

  Outputs (APK, ASK). APK is public and ASK is private to the attribute authority.
- **AttGen($ASK, \mathcal{A} \subset \mathbb{A}, \tau$).** Chooses random $r_a \leftarrow \mathbb{Z}_p^*$ for all $a \in \mathcal{A}$, and outputs

$$SK_{\mathcal{A},\tau} := \{(S_a, r_a)\}_{a \in \mathcal{A}} := \{(h^{(1/(b+c\cdot\tau||a+dr_a)}, r_a)\}_{a \in \mathcal{A}} \in (\mathbb{H} \times \mathbb{Z}_p^*)^{|\mathcal{A}|}.$$

$SK_{\mathcal{A},\tau}$ is privately given to the user associated to τ.

- **AttVer**$(APK, \mathcal{A} \subset \mathbb{A}, \tau, SK_{\mathcal{A},\tau})$. Parses $SK_{\mathcal{A},\tau}$ as $\{(S_a, r_a)\}_{a \in \mathcal{A}}$.

  Outputs 1 if $e(BC^{\tau \| a} D^{r_a}, S_a) = e(g, h)$ for all $a \in \mathcal{A}$, and 0 otherwise. Note that the algorithm can also be applied with a single attribute/signature pair.

- **Sign**$(PK, \psi, m, SK_{\mathcal{A},\tau})$. Assume that $\psi(\mathcal{A}) = 1$.

  Parses $SK_{\mathcal{A},\tau}$ as $\{\sigma_a := (S_a, r_a)\}_{a \in \mathcal{A}}$.

  Chooses random $\gamma \leftarrow \mathbb{Z}_p^*$, and computes $E_1 := \lambda^\gamma$, $E_2 := C^\tau \Lambda^\gamma$.

  Define $\tilde{\psi}(\cdot) = \psi(\cdot) \vee a_{m,\psi}$ where $a_{m,\psi} \in \mathbb{A}'$ is a pseudo-attribute.

  Let $\{a_1, \ldots, a_n\}$ denote the attributes that appear in the predicate $\tilde{\psi}$.

  For each i let VK_i be APK if $a_i \in \mathbb{A}$, or TVK otherwise.

  For each i let $\tilde{\sigma}_i$ be σ_{a_i} if $a_i \in \mathcal{A}$, or an arbitrary value otherwise.

  Let $\Phi[VK, m, \psi]$ denote the Boolean expression:

  $$\exists \tau, \sigma_1, \ldots, \sigma_n, \gamma : \tilde{\psi}(\{a_i : \text{AttVer}(VK_i, a_i, \tau, \sigma_i) = 1\}) = 1 \wedge$$

  $$\wedge\ E_2 C^{-\tau} = \Lambda^\gamma\ \wedge\ E_1 = \lambda^\gamma.$$

Computes the proof

$$\Sigma \leftarrow Groth.Prove(CRS; \Phi[VK, m, \psi]; (\tau, \{\tilde{\sigma}_i\}_{i=1}^n, \gamma), E_1, E_2).$$

Outputs $\sigma = (\Sigma, E_1, E_2)$ as the signature.

- **Veri**(PK, ψ, m, σ).   Computes $Groth.Veri(CRS; \Phi[VK, m, \psi]; \Sigma, E_1, E_2)$. Outputs the result.

- **Link**$(PK, (m', \psi', \sigma'), (m'', \psi'', \sigma''), mlk)$. Outputs 1 if $e(E_1', mlk) \cdot e(E_2'', h) = e(E_1'', mlk) \cdot e(E_2', h)$, and 0 otherwise.

This completes the description of our instantiation.

5   Correctness and Security Proofs of the ABS-CL Construction

Correctness of the signature follows from correctness of the first instantiation from [18]. We cover the additional details produced by adding the controllable linkability.

The main challenge in signing is to express the logic of the expression Φ in the Groth-Sahai system. We let $< Z >$ denote the formal variable corresponding to a commitment of the element Z in Groth-Sahai. Whether Z is in $\mathbb{G}$ or $\mathbb{H}$ will be clear from context.

If we take as a starting point the Groth-Sahai proof for the predicate in [18], to enhance this scheme with the desired functionality we must prove additionally, and simultaneously, that

$$\exists \gamma : (E_2 C^{-\tau}) = \Lambda^\gamma \wedge E_1 = \lambda^\gamma.$$

Since we want E_1 and E_2 to be public, the values h, λ, and Λ are already public, and we must keep secret γ and C^τ, then it suffices to commit to $< h^\gamma >$ and $< C^\tau >$ and prove in the Groth-Sahai system the following equations:

$$e(E_1, h) = e(\lambda, < h^\gamma >),\ \text{and}$$

$$e(E_2, h) = e(< C^\tau >, h) \cdot e(\Lambda, < h^\gamma >).$$

These equations prove that E_1 and E_2 have the *correct form*: first that within E_2 there is a factor C^τ, where C^τ is that used in the rest of the proof, and second that the remaining factor of E_2 is an exponentiation of Λ to the value $\xi = log_\lambda(E_1)$.

Correctness of the Link algorithm for honestly generated signatures is easily checked. Given two valid signatures created by users with tags τ' and τ'', we have that Link outputs 1 if and only if

$$e(C^{\tau'} \cdot \Lambda^{\gamma'}, h)e(\lambda^{\gamma'}, h^\xi)^{-1} = e(C^{\tau''} \cdot \Lambda^{\gamma''}, h)e(\lambda^{\gamma''}, h^\xi)^{-1},$$

$$e(C^{\tau'}, h)e(\lambda^{\xi\gamma'}, h)e(\lambda, h)^{-\gamma'\xi} = e(C^{\tau''}, h)e(\lambda^{\xi\gamma''}, h)e(\lambda, h)^{-\gamma''\xi},$$

$$e(C^{\tau'}, h) = e(C^{\tau''}, h),$$

which implies that $C^{\tau'} = C^{\tau''}$, and hence $\tau' = \tau''$, thus proving the link between the signatures.

We now show that our ABS-CL construction is secure under the notions of unforgeability and anonymity we propose. We state all of our security results, and then provide proof sketches for them. Complete proofs of our theorems will be available in the full version of the paper.

Theorem 9. *Signature-Unforgeable*

The proposed ABS-CL scheme is signature-unforgeable if the chosen NIZK scheme is sound, and the signature scheme used by the attribute authority is secure.

Proof. Assuming that there is an adversary F that can break the unforgeability of signatures of the ABS-CL scheme, we will construct an efficient algorithm S which violates, with comparable advantage, the security of the attribute authority's signature scheme, namely the Boneh-Boyen signature scheme. Given the verification key VK for the Boneh-Boyen scheme, the algorithm S runs one of two possible scenarios:

Scenario 1. Run (CRS, θ) and generate the rest of the parameters honestly. Give the resulting TPK and mlk to adversary F as the simulated result of TSetup. When F requests the signing keys of an attribute set, or the secret key of the attribute authority, compute the answer honestly and reply. If F makes a (m, ψ) query to the signing oracle, make an oracle query to the Boneh-Boyen oracle for the pseudo-attribute associated with (m, ψ). Utilize the response to construct a simulated ABS-CL signature and forward it to F.

When the adversary F outputs (m^*, ψ^*, σ^*), extract a witness using the trapdoor θ, which succeeds with overwhelming probability. This means that the extraction has obtained a signature for the pseudo-attribute associated to (m^*, ψ^*), or enough attributes to satisfy ψ^*. If it contains a signature for the pseudo-attribute, then it represents a forgery in the Boneh-Boyen scheme since S never queried (m^*, ψ^*).

Scenario 2. Use VK as the APK of the attribute authority. Setup CRS and all of the parameters corresponding to the trustee honestly, and give the resulting TPK and mlk to the adversary F. When the adversary requests the signing keys of an attribute set, forward the request to the Boneh-Boyen oracle. If F makes a (m, ψ) query to the signing oracle, then respond with a signature on the pseudo-attribute associated with (m, ψ).

When F finally outputs (m^*, ψ^*, σ^*), extract a witness. Similarly as in Scenario 1, it will succeed with overwhelming probability, and it will contain a signature for the pseudo-attribute associated to (m^*, ψ^*), or enough attributes to satisfy ψ^*. If it contains enough attributes, then at least one of them represents a forgery in the Boneh-Boyen scheme, since there is at least one non-corrupted attribute needed to satisfy the predicate ψ^*.

Observe that the output of F must be a forgery in one of the scenarios, and that from the view of F both are identical. Hence, if F has a non-negligible advantage ϵ in the ABS-CL game, and a ratio r of its successful outputs contain the pseudo-attribute, then S has an advantage of $\frac{1}{2}r\epsilon + \frac{1}{2}(1-r)\epsilon = \frac{1}{2}\epsilon$, which is non-negligible. $\qquad\square$

Theorem 10. *Link-Unforgeable*

The proposed ABS-CL scheme is link-unforgeable if the chosen NIZK scheme is sound, and the signature scheme used by the attribute authority is secure.

Proof. We show that given the structure of signatures in our scheme, if an adversary F can create a signature linking to an honest user, then the output must contain a forgery of a signature of said honest user, which contradicts the previous theorem. Let (m^*, ψ^*, σ^*) be the output of the adversary. Then $\sigma^* = (\Sigma^*, E_1^*, E_2^*)$ and $\sigma = (\Sigma, E_1, E_2)$, are such that $E_2^* = C^{\tau^*}\lambda^{\gamma^*\xi}$ and $E_2 = C^\tau \lambda^{\gamma\xi}$, for some values τ, τ^*, where τ is the identifier of an honest user. Since

$$\text{Link}(PK, (m^*, \psi^*, \sigma^*), (m, \psi, \sigma), mlk) = 1,$$

we have that $\text{Veri}(\psi^*, m^*, \sigma^*) = 1$, and that $C^\tau = C^{\tau^*}$, which can only be possible if $\tau = \tau^*$. Hence, σ^* constitutes a signature forgery of message m^* under predicate ψ^* for the honest user with identifier τ. $\qquad\square$

We now turn to anonymity, against adversaries with and without the linking key. Note that contrarily to [18] it is not possible for this scheme to have *unconditional* anonymity, since some information about the identity is leaked through the master linking key, which itself may be found with unbounded computational power. Indeed, recall that $mlk = h^\xi$, and that h, λ, and $\Lambda = \lambda^\xi$ are public values.

Theorem 11. *Anonymity (without the linking key)*

The given construction of an ABS-CL scheme is anonymous against an adversary without access to the master linking key, if the underlying proof system is secure, under the DDH assumption.

Proof. We show that if an adversary F wins the anonymity game in the ABS-CL, then we can construct at least one of the following adversaries: S_1 violating the zero-knowledge property, S_2 breaking the witness indistinguishability of the authority's signature scheme, and S_3 violating the DDH.

We have that real and simulated proofs of statements are indistinguishable. The adversary S_1 would get the CRS and set up the rest of the parameters honestly. The attribute authorities and queries to the linking oracle are answered honestly, while sign queries may be simulated. When the adversary F outputs the query to the challenge, which contains two possible sets of parameters for the generation of the challenge signature, none of which have been queried to the signing or linking oracles. Then S_1 chooses one of the sets and forwards it to the zero knowledge challenger, receiving back a proof Σ of the statement Φ, from which S_1 completes an ABS-CL signature and forwards it to F. Recall that the proof Σ may be real or simulated. In the case of a real proof, breaking anonymity can be reduced to an adversary S_2 violating the indistinguishability of the authority's signature scheme, through a series of games such as in [1]. If the proof is a simulated one, then we have a signature $\sigma = (\Sigma, E_1, E_2)$, where Σ is a simulated proof of the statement Φ, and E_1, E_2 are of the form $E_1 = \lambda^\gamma$ and $E_2 = C^\tau \Lambda^\zeta$ for some identifier τ and values $\gamma, \zeta \in \mathbb{Z}_p^*$. If the latter is used to win the indistinguishability game then an adversary S_3 can be built with an advantage against the DDH. □

Theorem 12. *Anonymity (with the linking key)*

The given construction of an ABS-CL scheme is anonymous against an honest but curious linker, with access to the master linking key, but not colluding with the attributes authorities, under the DDH assumption.

Proof. The challenge query must not contain identifiers queried to the signing oracle, which prevents the linker from doing a trivial link check on the challenge signature. Since the identifiers τ_i must belong to honest users, it means that the values τ_i and C^{τ_i} are unknown to the linker and thus he cannot use the bilinear check against $e(C^{\tau_i}, h)$. It is then straightforward to prove that the security reduces to that of the DDH. □

6 Conclusion

We have introduced the new feature of Controllable Linkability in the ABS setting, which allows a trusted entity in possession of the master linking key to determine if a pair of signatures were created by the same signer. We presented an instantiation of an ABS-CL scheme. Finally, we also prove that the constructed scheme achieves anonymity whether in possession or not of the master linking key, and unforgeability against signatures and links. It remains to extend the scheme to the multi-authority setting, which requires adjustments on the construction, the security notions and proofs.

Acknowledgements. The second author was partially funded by the Fonds National de la Recherche (FNR), Luxembourg. We would like to thank Ali El Kaafarani, and the anonymous reviewers of INTRUST 2015, for their valuable comments and helpful suggestions.

References

1. Bernhard, D., Fuchsbauer, G., Ghadafi, E., Smart, N.P., Warinschi, B.: Anonymous attestation with user-controlled linkability. Int. J. Inf. Secur. **12**(3), 219–249 (2013)
2. Bohli, J.M., Pashalidis, A.: Relations among privacy notions. ACM Trans. Inf. Syst. Secur. (TISSEC) **14**(1), 4 (2011)
3. Boneh, D., Boyen, X.: Short signatures without random oracles. In: Cachin, C., Camenisch, J.L. (eds.) EUROCRYPT 2004. LNCS, vol. 3027, pp. 56–73. Springer, Heidelberg (2004). http://dx.doi.org/10.1007/978-3-540-24676-3_4
4. Brickell, E., Camenisch, J., Chen, L.: Direct anonymous attestation. In: Proceedings of the 11th ACM Conference on Computer and Communications Security, CCS 2004, pp. 132–145. ACM, New York (2004). http://doi.acm.org/10.1145/1030083.1030103
5. Changxia, S., Wenping, M.: Secure attribute-based threshold signature without a trusted central authority. Spec. Issue Adv. Comput. Electron. Eng. **7**(12), 2899 (2012)
6. Chaum, D., van Heyst, E.: Group signatures. In: Davies, D.W. (ed.) EUROCRYPT 1991. LNCS, vol. 547, pp. 257–265. Springer, Heidelberg (1991). http://dx.doi.org/10.1007/3-540-46416-6_22
7. El Kaafarani, A., Chen, L., Ghadafi, E., Davenport, J.: Attribute-based signatures with user-controlled linkability. In: Gritzalis, D., Kiayias, A., Askoxylakis, I. (eds.) CANS 2014. LNCS, vol. 8813, pp. 256–269. Springer, Heidelberg (2014). http://dx.doi.org/10.1007/978-3-319-12280-9_17
8. El Kaafarani, A., Ghadafi, E., Khader, D.: Decentralized traceable attribute-based signatures. In: Benaloh, J. (ed.) CT-RSA 2014. LNCS, vol. 8366, pp. 327–348. Springer, Heidelberg (2014). http://dx.doi.org/10.1007/978-3-319-04852-9_17
9. Gagné, M., Narayan, S., Safavi-Naini, R.: Short pairing-efficient threshold-attribute-based signature. In: Abdalla, M., Lange, T. (eds.) Pairing 2012. LNCS, vol. 7708, pp. 295–313. Springer, Heidelberg (2013). http://dx.doi.org/10.1007/978-3-642-36334-4_19
10. Ghadafi, E.: Stronger security notions for decentralized traceable attribute-based signatures and more efficient constructions. In: Nyberg, K. (ed.) CT-RSA 2015. LNCS, vol. 9048, pp. 391–409. Springer, Heidelberg (2015). http://dx.doi.org/10.1007/978-3-319-16715-2_21
11. Groth, J., Sahai, A.: Efficient non-interactive proof systems for bilinear groups. In: Smart, N.P. (ed.) EUROCRYPT 2008. LNCS, vol. 4965, pp. 415–432. Springer, Heidelberg (2008). http://dx.doi.org/10.1007/978-3-540-78967-3_24

12. Herranz, J., Laguillaumie, F., Libert, B., Ràfols, C.: Short attribute-based signatures for threshold predicates. In: Dunkelman, O. (ed.) CT-RSA 2012. LNCS, vol. 7178, pp. 51–67. Springer, Heidelberg (2012). http://dx.doi.org/10.1007/978-3-642-27954-6_4

13. Hwang, J.Y., Chen, L., Cho, H.S., Nyang, D.: Short dynamic group signature scheme supporting controllable linkability. IEEE Trans. Inf. Forensics Secur. 10(6), 1109–1124 (2015)

14. Hwang, J.Y., Lee, S., Chung, B.H., Cho, H.S., Nyang, D.: Short group signatures with controllable linkability. In: 2011 Workshop on Lightweight Security Privacy: Devices, Protocols and Applications (LightSec), pp. 44–52, March 2011

15. Hwang, J.Y., Lee, S., Chung, B.H., Cho, H.S., Nyang, D.: Group signatures with controllable linkability for dynamic membership. Inf. Sci. 222, 761–778 (2013). http://www.sciencedirect.com/science/article/pii/S0020025512005373, including Special Section on New Trends in Ambient Intelligence and Bio-inspired Systems

16. Khader, D.: Attribute-based group signature with revocation (2007)

17. Maji, H.K., Prabhakaran, M., Rosulek, M.: Attribute-based signatures: achieving attribute-privacy and collusion-resistance (2008)

18. Maji, H.K., Prabhakaran, M., Rosulek, M.: Attribute-based signatures. In: Kiayias, A. (ed.) CT-RSA 2011. LNCS, vol. 6558, pp. 376–392. Springer, Heidelberg (2011). http://dx.doi.org/10.1007/978-3-642-19074-2_24

19. Okamoto, T., Takashima, K.: Efficient attribute-based signatures for non-monotone predicates in the standard model. In: Catalano, D., Fazio, N., Gennaro, R., Nicolosi, A. (eds.) PKC 2011. LNCS, vol. 6571, pp. 35–52. Springer, Heidelberg (2011). http://dx.doi.org/10.1007/978-3-642-19379-8_3

20. Okamoto, T., Takashima, K.: Decentralized attribute-based signatures. In: Kurosawa, K., Hanaoka, G. (eds.) PKC 2013. LNCS, vol. 7778, pp. 125–142. Springer, Heidelberg (2013). http://dx.doi.org/10.1007/978-3-642-36362-7_9

21. Park, H., Kent, S.: Traceable anonymous certificate (2009)

22. Shahandashti, S.F., Safavi-Naini, R.: Threshold attribute-based signatures and their application to anonymous credential systems. In: Preneel, B. (ed.) AFRICACRYPT 2009. LNCS, vol. 5580, pp. 198–216. Springer, Heidelberg (2009). http://dx.doi.org/10.1007/978-3-642-02384-2_13

23. Tate, S.R., Vishwanathan, R.: Expiration and revocation of keys for attribute-based signatures. In: Samarati, P. (ed.) DBSec 2015. LNCS, vol. 9149, pp. 153–169. Springer, Heidelberg (2015). http://dx.doi.org/10.1007/978-3-319-20810-7_10

A Causality-Based Model for Describing the Trustworthiness of a Computing Device

Jiun Yi Yap[✉] and Allan Tomlinson

Information Security Group, Royal Holloway, University of London,
Egham, Surrey TW20 0EX, UK
Jiun.Yap.2012@live.rhul.ac.uk, Allan.Tomlinson@rhul.ac.uk

Abstract. The ability to describe the trustworthiness of a computing device is an important part of the process to establish end-to-end trust. With the understanding that the trustworthiness of a computing device relies on its capabilities, we report on and contribute a novel causality-based model. This causality-based model represents information about the dependencies between trust notions, capabilities, computing mechanisms and their configurations. In this work, the concept of causality within the model is defined first. This involves detailing the semantic meaning of the terms used in the model. A pictorial representation is then developed to show the causal dependencies as a graph. This step specifies the vertices and edges used in the causal graph. To implement the causality-based model, the causal graph was translated into an eXtensible Markup Language schema and added to the Metadata Access Point database server of the Trusted Network Connect open architecture. Finally, the trust assessment of the causal graph is explained.

1 Introduction

Modern computing devices are diverse and interconnected by dynamic and heterogeneous networks. In the National Cyber Leap Year Summit, the participating researchers reported that end-to-end trust will be a game changing technology when deployed in this type of computing environment [1]. A key component to building end-to-end trust is the ability to describe the trustworthiness of a computing device. The benefit of this is that any other computing device can select its mode of participation in a computer network according to the level of trustworthiness offered by the corresponding computing device. Similarly, Grawrock et al. explained that there is a need for computing devices to communicate their trustworthiness so that the parties involved can understand and manage security risks [2].

There are two facets to this ability. First of all, it is the description of the capabilities of a computing device that could give rise to its trustworthiness. This requires marking up this description with metadata in a well understood and consistent manner. Such annotation will enable the description of the capabilities to be processed by a computer [14]. The second facet refers to evidence that conveys assurance in certain capabilities of the computing device. A prevailing technique is the use of attestation which vouches for the identity and state of

© Springer International Publishing Switzerland 2016
M. Yung et al. (Eds.): INTRUST 2015, LNCS 9565, pp. 130–149, 2016.
DOI: 10.1007/978-3-319-31550-8_9

a computing device's software stack. In this paper, we take the first step towards describing the capabilities of a computing device.

Our central contribution is a novel causality-based model for describing the capabilities of a computing device that give rise to its trustworthiness. As far as we know, this is the first attempt at this challenge. The main advantage of this approach is that the description captures the causal dependencies between trust notions, capabilities, mechanisms and configurations, and this information is useful for intelligent processing. Moreover, it uses a graph to describe a computing device at various levels of abstraction in a way that is easily comprehended. Lastly, the causal graph can be made machine-readable by translating it into a format that uses markup language, such as the eXtensible Markup Language (XML).

The rest of this paper is organized as follows. Section 2 of this paper gives the background to the causality-based approach. Section 3 defines the concept of causality for this model and make clear the semantics of the terms used. Section 4 defines the graph and illustrates how it is used to show the dependencies between trust notions, capabilities, computing mechanisms and their configuration. This is followed by Sect. 5 that describes how the graph can be implemented as a XML schema and we discuss how this schema was applied to the Metadata Access Point (MAP) database server of the Trusted Network Connect (TNC) open architecture. Section 6 explains how we can carry out trust assessment. Section 7 discusses future work and Sect. 8 reviews related works. This paper concludes in Sect. 9.

2   A Causality-Based Approach

The design of the model aims to meet the following requirements:

- To describe the capabilities of a computing device that could give rise to its trustworthiness.
- To define the model in a clear and easy to understand manner.
- To support the digital representation of this model by translating it into a machine-readable format.

The intention of these requirements are to guide the development of the model and make sure that it can be implemented in practice. It is not the intention of this paper to specify how the descriptions are created and updated. We can assume that such descriptions are created by the designer of the computing device and the descriptions are updated by agents installed on that computing device.

To describe the capabilities of a computing device that give rise to its trustworthiness, we refer to the technical models described in the National Institute of Standards and Technology (NIST) Special Publication 800-33 [3]. These models encompass information at several levels. They range from high level security notions to low level specific technical details. For example, the low level technical mechanism of cryptographic key management is described as enabling the capability of access control enforcement which in turn supports the security notion of confidentiality. The NIST publication is intended to describe the technical foundations that underlie security capabilities. Since our intentions are broadly aligned, we will frame our description of the capabilities of a computing device on the structure of the technical models described in the NIST publication. The description will cover the trust notions, capabilities, mechanisms and configurations of a computing device. We also note that if any of the technical foundations are missing, the resulting high level capability and notion will not exist. This observation is interpreted as causality.

A general causal model consists of a set of equations of the form

$$x_i = f_i(pa_i, u_i), \ i = 1,...,n,$$

where pa_i stands for the set of variables that directly determine the value of X_i and where U_i represents errors or disturbances due to omitted factors [4]. This functional relationship can be thought of as Laplace's quasi-deterministic concept of causality. Bayesian networks are usually used to represent this general causal model and, for example, U_i can be used to indicate the probability of a causal dependency when affected by factors such as an attack on the computing device. However, this approach is sophisticated and at this stage, it is beyond our design requirements. Hence, we decided to set U_i to zero (i.e. no errors) in our causality-based model. By setting U_i to zero, the causal model loses the ability to deal with the probability of a causal dependency. As our primary concern is to introduce a model to describe the causal dependencies between trust notions, capabilities, computing mechanisms and their configuration and that this model has to work in practice, missing the above ability does not affect the description. Nevertheless, we will discuss more about setting U_i to non-zero in Sect 7.

With this understanding of a causal model, we can say that: *A cause is defined to be an object followed by another, where, if the first object had not been, the second would never had existed* [5]. This concept of causality has important applications in computer science, such as intelligent planning and processing [6]. Whenever we seek to explain a set of computations that unfold in a specific scenario, the explanation produced must address the cause and effect of these computations. The generation of such explanations requires the analysis of the concept of causality and the development of a data model to characterize the account.

On the other hand, the practical application of causality requires it to transform into a graph that is founded on mathematics and logic [4]. A graph consists of a set V of vertices and a set E of edges. The vertices in this graph represent the variables and the edges denote a certain causal relationship holds in pairs of variables. As a causal explanation describes how a variable is caused by another variable, this description of dependency path can be written as a triplet (v1, e, v2) where v1, v2 $\in V$ and e $\in E$. In other words, a source variable v1 is related to destination variable v2 through the causal dependency e. Consequently, this can also be expressed as a graph where e is an edge from source vertex v1 to destination vertex v2. Hence, we can conjecture that a causal graph composes of multiple triplets and they form a larger graph with numerous vertices that are connected by edges.

With this background knowledge, we can envisage that this graph will have a data structure that depicts the causal dependences between the low level technical mechanisms and high level capabilities and notions. Although it can be argued that such information can be obtained from various sources, the main advantage of this model is that the information about technical mechanisms, configuration, capabilities and trust notions are linked together meaningfully by their causal dependencies. As a result, another party can carry out intelligent processing on this information and decide how it will interact with this computing device.

3   Basic Definitions

In this section, we will define the semantic meaning of the terms used in this causality-based model. As the causality-based model has to be transformed into a graph for practical application, the following Definitions 2 to 5 will refer to the kind of vertices in the graph.

Definition 1 (Causality). We refer to the definition of causality in the previous section and define causality in this model to be the use of a configured mechanism, or a set of configured mechanisms, and if one of the configured mechanism is not used, the capability and the resulting trust notion that arise out of the use of the configured mechanisms will not exist. For example, the trust notion of confidentiality relies on the capability of disk encryption which in turn is derived from the usage of a symmetric cryptographic mechanism.

We then produce the following definitions for the key terms in Definition 1.

Definition 2 (Mechanism). It is a computation process that has at least one input and at least one output. Although how the computation works is defined by computer code, the quality of the output can be influenced by the applied configuration. For example, a symmetric encryption mechanism is configured to use a symmetric key of certain size, then takes in data and produces the encrypted form of that data.

Definition 3 (Configuration). A set of parameters that affect the output of a mechanism. For example, the key size of an AES symmetric encryption mechanism needs to be specified as different applications require different key sizes.

Definition 4 (Trust Notion). A notion reflects a specific behavior of a computing system. We understand that there is a large body of research on trust notions. Thus, to focus our research effort, we scoped this work to the trust notions described our previous study [15]. These trust notions are confidentiality, integrity, identity, authenticity, availability and expected behaviour. These are objective notions that are related to technical properties of a computing device.

Definition 5 (Capability). A capability can be considered as a high level description of a mechanism or a set of cooperating mechanisms. It refers to the ability to perform certain tasks. A capability of one computing system is the same as another system if this capability is derived from the same set of mechanisms and configurations.

Various notions of causal dependencies were considered for the causality-based model. A strong notion of causal dependency would provide a detailed explanation of how an effect is caused. However, such strong notion of causal dependency was not practical as one could argue that additional factors may have influenced the outcome. For example, if the computation has been occurring on hardware that is operating within its allowed temperature range. Therefore, we decided that weaker notions which describe only the core meaning of the causal dependencies would be more suitable. This decision is supported by two considerations:

- *Usability.* We expect that the description of the capabilities could be produced without detailed knowledge of how the mechanisms interact and how the configuration affects the behaviour of the mechanisms. Thus, weaker notions allow the causality-based model to be used in practice by non experts.
- *Composability.* We desire that multiple causal graph can be combined to reflect more complex capabilities. This is true in practical application whereby a complex capability can be composed of various mechanisms spread across diverse computing systems. Hence, weaker notions allow a more flexible interpretation of the causal dependencies and avoid complications due to contrasting explanation used by different computing systems.

Nevertheless, stronger notions of causality for specific applications, can be developed as subclasses to the dependencies defined in our model. While the above definitions refer to the kind of vertices in a causal graph, the edges between the nodes will represent their causal dependencies. On this point, we propose the following causal dependencies for this model:

Definition 6 (Mechanism *CallsOn* Mechanism). A mechanism can call on another mechanism during its computation process. This relation can be one to one, one to many, many to one or many to many. However, this causal dependency is affected by the configuration of the mechanism. A mechanism with the same configuration can be called on multiple times by other mechanisms. If this mechanism has another configuration, then it must be represented again on the causal graph. A calling mechanism can only complete its computation process when the mechanism that it called on has completed its computation process. For example, a software application calls on a key generator and a symmetric encryption mechanism when it is performing disk encryption.

Definition 7 (Capability *DerivesFrom* Mechanism). A capability is derived from a mechanism or a set of cooperating mechanisms. If a capability is derived from a set of cooperating mechanisms, all these mechanisms must have begun and completed their computation processes before the capability can exist. The same constraint applies to the situation when a capability is derived from one mechanism. The quality of a capability will be affected if a mechanism or a configuration is not from a specified set. For example, the capability of disk encryption is derived from a software application that encrypts data using symmetric cryptography and manages the key used in the encryption process.

Definition 8 (Trust Notion *ReliesOn* Capability). The existence of a trust notion relies on a capability or a set of capabilities. The mechanisms that the capability depends on must have completed their computation process before the trust notion can exist. A trust notion can rely on more than one capability. For example, the trust notion of confidentiality relies on the capability of disk encryption. The same trust notion of confidentiality can also rely on the capability of network encryption which uses a different set of mechanisms from the capability of disk encryption.

Definition 9 (Mechanism *Uses* Configuration). Each mechanism has a maximum of one configuration for each account of causality. In other words, it is a one to one relation. If the same mechanism uses more than one configuration, then the mechanism shall be represented again with another configuration.

We have introduced the definition of the terms used in the causality-based model. However, there may be ambiguity in the definition and hence we introduce set-theoretic definitions in Fig. 1 to further clarify the causality-based model.

In Fig. 1, lines 1 to 4 define the term mechanism, configuration, trust notion and capability as sets. These terms correspond to vertices when transformed to a graph. Line 5 further defines that M is a subset of MECHANISM and element m only belongs to if and only if m has a property P. P refers to the property that this set of mechanisms work together to give rise to a capability. Line 6 says that CF is a subset of CONFIGURATION and that element cf belongs to CF if and only if cf has a property Q. Q refers to the property that this set of configurations are applicable to a particular set of mechanisms that give rise to a capability.

Lines 7 and 8 are the interpretation of the definition of causality in this model. It says that there is a function f such that it maps elements of the TRUSTNOTION to the set CAPABILITY. Then there is a complex function g such that g(CAPABILITY) produces a set that contains the mapping of M to CF. Lines 9 to 12 refer to the edges that link the vertices. *CallsOn, ReliesOn*

1. [MECHANISM] the set of all possible mechanisms in computing systems.
2. [CONFIGURATION] the set of all possible configurations in computing systems.
3. [TRUSTNOTION] the set of all possible trust notions enabled by computing systems.
4. [CAPABILITY] the set of all possible capabilities provided by computing systems.

5. M is a subset of MECHANISM such that the elements of M are mechanisms that work together to give rise to a capability and trust notion. Therefore,
   m $\in$ MECHANISM
   M = { m | P(m) }
6. CF is a subset of CONFIGURATION and elements of CF are configurations for a particular set of mechanisms. Therefore,
   cf $\in$ CONFIGURATION
   CF = { cf | Q(cf) }
   Following definition 1, we have:
7. f : TRUSTNOTION $\rightarrow$ CAPABILITY
8. g : CAPABILITY $\rightarrow$ (M $\rightarrowtail$ CF)

9. CallsOn : MECHANISM $\rightarrow$ MECHANISM
10. ReliesOn : TRUSTNOTION $\rightarrow$ CAPABILITY
11. Uses : MECHANISM $\rightarrowtail$ CONFIGURATION
12. DerivesFrom : CAPABILITY $\rightarrow$ MECHANISM
13. $\forall$ tn:TRUSTNOTION, m:MECHANISM, cf:CONFIGURATION $\bullet$ tn R^+ m = $\emptyset$
    $\land$ tn R^+ cf = $\emptyset$
14. $\forall$ cp:CAPABILITY, cf:CONFIGURATION $\bullet$ cp R^+ cf = $\emptyset$

Fig. 1. Definition of causality-based model.

and *DerivesFrom* refers to a one to one or one to many mapping. For *Uses*, we use the symbol for an partial injective function to say that each mechanism has only one configuration at a time and there may exist a mechanism that does not need any configuration. These edges shall be directed and the direction is from the range to the domain of the functions CallsOn, ReliesOn, DerivesFrom and Uses. The way these edges are directed reflects our requirement to describe the causal dependencies between trust notions, capabilities, mechanisms and configurations. This is a mapping from abstract concepts to low level technical primitives. Meanwhile, the directed edge ReliesOn is the manifestation of function f in line 7 while the directed edge DerivesFrom is the manifestation of function g in line 8. Finally, lines 13 and 14 clarify that transitive closure is not allowed. Particularly, it addresses the constraint that neither trust notion nor capability can arise only out of configuration. The other reason is that we want to capture all the causal dependencies.

4   Graph Definition

This section will define the graph that is required for the practical application of the causality-based model. We named this as the causal graph. The following defines this causal graph.

1. Vertices of the causal graph are the elements of MECHANISM, CONFIG-URATION, CAPABILITY and TRUSTNOTION and they are given unique identifiers. Two elements from the same set are the same if they have the same identifiers.
2. Elements of MECHANISM, CONFIGURATION and CAPABILITY can be tagged to a computing system. This supports composability; i.e. a capability can be derived from mechanisms held in more than one computing system. This also supports the situation where multiple capabilities from different systems give rise to a trust notion.
3. Edges of the causal graph are identified by the vertices they connect. Vertices are elements of the sets MECHANISM, CONFIGURATION, CAPABILITY and TRUSTNOTION and they have unique identifiers.
4. A causal graph is a set of vertices and edges as specified in this paper.
5. A proper causal graph contains a ReliesOn and a DerivesFrom edge, In other words, it explains the existence of a trust notion and capability. This ensures that a causal graph captures a valid causal dependency between the high level trust notion or capability and a low level technical mechanism. A proper causal graph is also acyclic due to the direction of the edges. It is possible that with additional definition of edges, a causal graph can be made cyclic but this will not be discussed in this paper.

Figure 2 provides a set-theoretic definition of the causal graph. Lines 1 to 5 declare the sets of identifiers. Lines 6 to 8 says that the vertices Mechanism, Configuration and Capability have unique identifiers and are associated with at least one computing system. We do not link Trust Notion to a particular computing system as it refers to a universal quality. The set for edges CallsOn, ReliesOn, Uses and DerivesFrom are specified from line 10 to 13. The information about the vertices and edges will be expressed as a triplet (v1, e, v2) where vertex v1 is related to vertex v2 through a causal dependency e. Line 14 says that a causal graph is a set containing the triples that describe the causal dependencies.

Lines 15 to 18 attempt to explain systematically how two graphs can be equal. Line 15 looks at the most basic causal dependency between a mechanism and a configuration. It says that their Uses edges are the same if the corresponding mechanisms and configurations are the same although their systems may not be the same. We then proceed to line 16 that defines how two CallsOn edges are the same. It says that the CallsOn edges between 2 separate sets mechanisms are the same if the elemental mechanisms from each set are the same and they use the same configurations although their system are not the same. Lines 17 and 18

1. MechanismId : primitive set
2. ConfigurationId : primitive set
3. TrustNotionId : primitive set
4. CapabilityId : primitive set
5. System : primitive set
6. Mechanism : MechanismId $\rightarrow$ $\mathbb{P}$(System)
7. Configuration : ConfigurationId $\rightarrow$ $\mathbb{P}$(System)
8. Capability : CapabilityId $\rightarrow$ $\mathbb{P}$(System)
9. TrustNotion : TrustNotionId

10. CallsOn = Mechanism $\times$ Mechanism
11. ReliesOn = TrustNotion $\times$ Capability
12. Uses = Mechanism $\times$ Configuration
13. DerivesFrom = Capability $\times$ Mechanism
14. CausalGraph = $\mathbb{P}$(CallsOn) $\times$ $\mathbb{P}$(ReliesOn) $\times$ $\mathbb{P}$(Uses) $\times$ $\mathbb{P}$(DerivesFrom)

15. m1,m2 $\in$ Mechanism, cf1, cf2 $\in$ Configuration, s1, s2 $\in$ System.
    $\exists$ u1, u2 $\in$ Uses | u1 = ((m1,s1),(cf1, s1)), u2 = ((m2, s2), (cf2, s2)) $\bullet$ u1 = u2 $\Leftrightarrow$ (m1 = m2) $\wedge$ (cf1 = cf2) $\wedge$ ((s1 = s1) $\vee$ (s1 $\neq$ s2))
16. m1,m2, m3, m4 $\in$ Mechanism, u1, u2, u3, u4 $\in$ Uses, cf1, cf2, cf3, cf4 $\in$ Configuration, s1, s2 $\in$ System.
    $\exists$ co1,co2 $\in$ CallsOn | co1 = ((m1,s1),(m2,s1)), co2 = ((m3,s2), (m4, s2)), u1 = ((m1, s1),(cf1, s1)), u2 = ((m2, s1), (cf2, s1)), u3 = ((m3, s2),(cf3, s2)), u4 = ((m4, s2), (cf4, s2)) $\bullet$ co1 = co2 $\Leftrightarrow$ (m1 = m3) $\wedge$ (m2 = m4) $\wedge$ (cf1 = cf3) $\wedge$ (cf2 = cf4) $\wedge$ ((s1 = s1) $\vee$ (s1 $\neq$ s2))
17. DF1, DF2 $\subseteq$ DerivesFrom, M10, M11, M20, M21 $\subseteq$ Mechanism, CO1, CO2 $\subseteq$ CallsOn, U10, U11, U20, U21 $\subseteq$ Uses, CF10, CF11, CF20, CF21 $\subseteq$ Configuration.
    $\exists$ cp1, cp2 $\in$ Capability | DF1 = (cp1, M10), DF2 = (cp2, M20), CO1 = (M10, M11), CO2 = (M20, M21), U10 = (M10, CF10), U20 = (M20, CF20), U11 = (M11, CF11), U21 = (M21, CF21) $\bullet$ cp1 = cp2 $\Leftrightarrow$ (U10 = U20) $\wedge$ (U11 = U21) $\wedge$ (CO1 = CO2)
18. RO1, RO2 $\subseteq$ ReliesOn, CP1, CP2 $\subseteq$ Capability, DF1, DF2 $\subseteq$ DerivesFrom, M10, M11, M20, M21 $\subseteq$ Mechanism, CO1, CO2 $\subseteq$ CallsOn, U10, U11, U20, U21 $\subseteq$ Uses, CF10, CF11, CF20, CF21 $\subseteq$ Configuration.
    $\exists$ tn1, tn2 $\in$ TrustNotion | RO1 = (tn1, CP1), RO2 = (tn2, CP2), DF1 = (CP1, M10), DF2 = (CP2, M20), CO1 = (M10, M11), CO2 = (M20, M21), U10 = (M10, CF10), U20 = (M20, CF20), U11 = (M11, CF11), U21 = (M21, CF21) $\bullet$ tn1 = tn2 $\Leftrightarrow$ (DF1 = DF2) $\wedge$ (CO1 = CO2) $\wedge$ (U10 = U20) $\wedge$ (U11 = U21)

Fig. 2. Defining the causal graph.

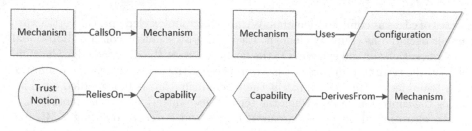

Fig. 3. Pictorial representation of the vertices and edges in the causal graph model.

make use of the definitions in lines 15 and 16. They explain how two capabilities or two trust notions can be the same. Line 17 says that two capabilities can be the same if they have the same set of Uses and same set of CallsOn. In other words, two capabilities are the same if they are derived from the same set of mechanisms and these set of mechanisms have the same set of configurations. Line 18 says that two trust notions can be the same if they rely on the same capabilities, and these capabilities are derived from the same mechanisms, and these mechanisms use the same configurations. Note that the requirement that the system may or may not be the same is implicit.

At this point, we will introduce the basic pictorial representation of vertices and directed edges of the causal graph. Mechanisms are represented as rectangles while configurations are trapeziums. Finally, trust notions are circles and capabilities are hexagons. Figure 3 shows this pictorial representation.

5   Praxis

We implemented the causality-based model to the Metadata Access Point (MAP) server of the Trusted Network Connect (TNC) open architecture. TNC is specified by the Trusted Computing Group and it enables the application and enforcement of security requirements for endpoints connecting to an enterprise network [7]. The MAP server acts as a database where metadata that describes endpoints is published. Security devices can subscribe to the MAP server and read the published metadata as part of a security process such as access control. Communications with the MAP server are carried out over the Interface for Metadata Access Point (IF-MAP) protocol. Standard sets of metadata are defined for use by the MAP server to determine information about an endpoint such as device status, location and characteristics. There are currently two standards for metadata; the IF-MAP Metadata for Network Security [8] and IF-MAP Metadata for Industrial Control System [9]. The specification of the standards defines the structure and content of the metadata and includes XML schemas to represent this information. We created a new XML schema to represent our causality-based model. This new XML schema is presented in Fig. 6 of the Appendix. This new XML schema declares the

4 types of edges as complex elements. Within the complex element, the vertices are further declared as complex element. Each complex element contains additional elements such as id and system. On the declaration for the directed edge CallsOn, we used the term "MainMechanism" to represent the calling mechanism and the term "SubMechanism" to represent the mechanism that is being called on. This avoids the confusion caused when the term "Mechanism" is not differentiated.

We also crafted an example causal graph and its corresponding XML format that is based on our schema. This causal graph is shown as a graph in Fig. 4 and as in XML format in Fig. 7 of the Appendix. As the causality-based model is a novel approach, we could not find any documentation that captures the causal dependencies of trust notions, capabilities, mechanisms and configurations. Thus, this example approximates a design of data storage encryption using the Trusted Platform Module (TPM) version 2.0 [10]. It describes how the capability of disk encryption is derived from a software application named "DiskLocker". This software application is represented as a mechanism and it calls on various TPM 2.0 subsystems. These TPM 2.0 subsystems are represented as mechanisms. The "DiskLocker" software and some of the TPM 2.0 subsystem uses configurations. For example, the configuration of "DiskLocker" state where the symmetric cryptographic key will be stored. In another example, the random number generator has to be configured with a seed value.

In this example, the "DiskLocker" software uses symmetric cryptography to encrypt data. The symmetric cryptographic mechanism is provided by the TPM 2.0. Before, the symmetric cryptographic mechanism of TPM 2.0 can be used, "DiskLocker" obtained a random number from TPM 2.0 random number generator mechanism. This random number is passed on to TPM 2.0 key generation mechanism which produces a symmetric cryptographic key. This symmetric cryptographic key is then used with the symmetric cryptographic mechanism to encrypt data. To protect the symmetric cryptographic key, "DiskLocker" stores it in TPM 2.0 using the TPM 2.0 non-volatile memory mechanism. The quality of the disk encryption capability is affected if "DiskLocker" uses another random number generator that is not trusted. As a result, the trust notion of confidentiality that relies on the capability of disk encryption will be weaker than the situation where a trusted random number is used. Although the actual operation of such an implementation is more complex than that is described here, we decided to show just the necessary causal dependencies to ease practical application. Nevertheless, the causality-based model can be expanded to include more precise causal dependencies.

On the identity of mechanisms and configurations, a straightforward way to represent them is to refer to the Common Platform Enumeration (CPE) Reference [11] and the Common Configuration Enumeration (CCE) Reference [12]) of the National Institute of Standards and Technology. CPE is an organised naming scheme for computing systems, software, and packages. It is based on Uniform Resource Identifiers (URI) and contains a formal name format that support name verification. CCE gives unique identifiers to system configurations to

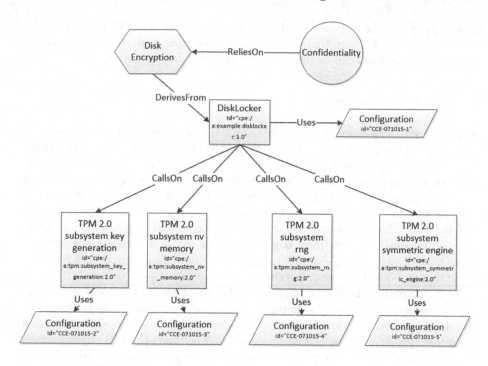

Fig. 4. Pictorial representation of a sample causal graph.

support correlation of configuration data. Since the example is an approxima-
tion, we assign dummy identifiers to support the simulation. We note that the
TPM 2.0 is neither listed in CPE nor CCE. Although the TPM is typically
implemented as one device and the mechanisms described in this example are
subsystems of the TPM device. However, the TPM specification allows the exact
implementation to vary. For example, only the SM4 symmetric cryptographic
algorithm is allowed in certain geographical regions. Therefore, we identified the
TPM subsystems as separate mechanisms and consequently they have their own
configurations.

We worked with the Fedora 20 operating system running the Linux kernel version 3.19.8. For the MAP server, we reviewed the open source *irond* version 0.5.6
MAP server developed by the trusted computing research group at the Hochschule
Hannover, University of Applied Sciences and Arts [13]. The latest version that
supported IF-MAP version 2.2 specification was used. It was a java program and
refered to stored XML schemas. We added our schema to that of IF-MAP Metadata for Network Security. Then we examined the open source *ifmapj* version 2.3.0
java library developed by the same team. We ran a MAP client program that made
use of this java library. This MAP client program encapsulates our example XML

shown in Fig. 7 of the Appendix and publishes it to the irond MAP server through an internal network. From this simulation, we managed to demonstrate how the causality-based model can be implemented on the TNC open architecture.

6   Trust Assessment

We developed an assessment rule to determine if a causal graph meets the requirement of a trust policy. A trust policy will consist of a set of assessment rules that lay out what are the identity and type of the vertices and edges required in a causal graph of a computing device before it can be trusted. We observed that the root of causal graphs will always be the vertices of Trust Notion. Thus, we can craft trust policies by starting with trust notions. To support this assessment, we formulated a rule that is based on determining the existence of a particular triplet in the causal graph. This is the most basic assessment rule and a trust policy will consist of a set of such basic assessment rules. This basic assessment rule is specified in the Backus Naus Form and it is shown in Fig. 5 below.

The specification begins by explaining the terms used. Line 3 states that the type of vertex can be either a mechanism, configuration, capability or trust notion. Line 4 then says that a target vertex is associated to an identity, and a system if the vertex type is a mechanism or a capability. The type of causal dependency is declared in line 5. Line 6 says that a triplet is projected as (SourceVertex, DependencyType, DestinationVertex). Hence, queries will be formulated around the concept of a triplet. Line 7 states a precise question on the existence of a destination vertex in a triplet. This question is the foundation of the assessment rule.

To develop a trust policy, we will have to understand how a mechanism interacts with another and what capability and trust notion does it enable. The trust policy will consist of numerous basic assessment rules that transverse through a causal graph and checking every branch from the root node to the leaf node. Figure 8 of appendix shows an example of a trust policy for the causal graph in Fig. 4.

1. (*U, RO, CO and DF are the short form for Uses, ReliesOn, CallsOn and Derives-From)
2. (*ME, CF, TN and CP are the short form for Mechanism, Configuration, Trust-Notion and Capability*)
3. <VertexType> ::= <ME> | <CF> | <CP> | <TN>
4. <SourceVertex>, <DestinationVertex> ::= <VertexType> <VertexID> | <VertexType> <id> <system>
5. <DependencyType> ::= <U> | <RO> | <CO> | <DF>
6. <Triplet>   ::=   "("<SourceVertex>","   <DependencyType>"," <DestinationVertex>")"
7. <BasicAssessmentRule> ::= "Is" <DestinationVertex> "∈ of a triplet containing ("<SourceVertex> "," <DependencyType>")?"

Fig. 5. Specification of the basic assessment rule.

The most direct way to assess a causal graph for trustworthiness is to process all the rules with a Boolean AND function. If the causal graph does not satisfy one rule, then the trust assessment will fail. However, we acknowledge that this assessment method is not flexible. If the trust assessment is to consider a range of values, then the rules can be assessed according to a Boolean Decision List [16]. However, we have to be careful with the complexity of the assessment process although we can induce that the problem space is achievable in polynomial time if the number of rules is finite.

To implement the basic assessment rule, we investigated the use of the XQuery language. For example, if the query is "Is SubMechanism id="cpe:/a:tpm:subsystem_key_generation:2.0"system="PHD_MC355_004"∈ of a triplet containing (MainMechanism id="cpe:/a:example:disklocker:1.0"system="PHD_MC355_004", CO)", you can issue the XQuery command in Fig. 9 of the Appendix. The return result will be "yes" in this example.

7   Future Work

Our causality-based model at this stage lays the foundation for two types of future work. The first type refers to the concept of predictions in causal models. For example, an evaluator can predict the trustworthiness of a computing device if it is presented with causal graphs describing the trust notions and capabilities. This type of future work will be a progression from our original intent of developing the causality-based model for end-to-end trust. The second type refers to the concept of interventions in causal models. For example, a security engineer can examine reference causal graphs and adjust either the mechanisms or configurations of a computing device to make sure that certain trust notions and capabilities are achieved. These two types of future work will require the development of algorithms that could understand the semantics of the causal graph and subsequently carry out intelligent processing.

We mentioned in Sect. 2 that our causality-based model does not have the ability to deal with the probability of a causal dependency. However, if we expand this causality-based model to include the probability of a causal dependency, then we can develop sophisticated models that predict the trustworthiness of a computing device in an uncertain environment. For example, we can model a weak causal dependency to reflect an attack and find out how the trust notions and capabilities are affected. This ability can be used in a cyber test range where it requires the modelling of computing devices under attack. This future work will require the use of bayesian networks that use probabilistic and statistical models.

8   Related Works

As far as we know, our work on the causality-based model is the first attempt at describing the capabilities of a computing device. The use of metadata for description can also be used in security ontologies. Kim et al. presented the Naval Research Laboratory (NRL) Security Ontology which focuses on the annotation of security mechanisms, protocols, algorithms, credentials and objectives [14]. The NRL Security Ontology consists of the core class of Security Concept. It has subclasses of Security Protocol, Security Mechanism and Security Policy. The Security Concept class is defined to support the Security Objective class. The other classes of the NRL Security Ontology include Security Algorithms, Security Assurance, Credentials, Service Security, Agent Security and Information Object. The authors explained that the classes of Service Security, Agent Security and Information Object classes are extensions of the DAML Security Ontology. On the other hand, the Credentials, Security Algorithm and Security Assurance classes provide values for properties defined for concepts in the Security Concept class. The Information Object class was added to allow for the annotation of web service inputs and outputs. Although the ontology based description was suitable for describing the make up of a computering device, the advantage of a causality based approach over an ontology is that the description concerns causal relationship and this is more useful for intelligent processing than the class hierarchy relationship presented by an ontology. In addition, the causality-based model is more flexible at addressing a wider range of applications where technical mechanisms could not always be represented in a class structure.

9   Conclusion

We revisit the requirements mentioned in Sect. 2. On the requirement to describe the capabilities of a computing device that give rise to its trustworthiness, we have developed a causality-based model that describe the causal dependencies between trust notions, capabilities, mechanisms and configurations. To make the model clear and easy to understand, we have given the basic definitions of the model and used set theory to further clarify the definitions. We then transformed the causality-based model into a causal graph for the purpose of data representation. Additional definitions are given for the graph model to show clearly how it can be used. During implementation, we have gained insights into how the causality-based model can be extended to the schemas used by the MAP server of the TNC open architecture and gave an example of a causal graph and its corresponding XML format. We also explained how to carry out trust assessment of the XML based causal graph. These practical exercises show that the requirement to support digital representation of this model is met.

A   Appendix

```xml
<?xml version="1.0"?>
<xsd:schema targetNamespace="causal_graph"
xmlns:xsd="http://www.w3.org/2001/XMLSchema">
<xsd:element name="causal_graph_data">
<xsd:complexType>
<xsd:choice maxOccurs="unbounded">
<xsd:element name="causal_graph_id" type="xsd:string"/>

<!--CallsOn-->
<xsd:element name="CallsOn">
<xsd:complexType>
<xsd:sequence>
<xsd:element name="MainMechanism">
<xsd:complexType>
<xsd:sequence>
<xsd:element name="id" type="xsd:string"/>
<xsd:element name="system" type="xsd:string"/>
</xsd:sequence>
</xsd:complexType>
</xsd:element>
<xsd:element name="SubMechanism" maxOccurs="unbounded">
<xsd:complexType>
<xsd:sequence>
<xsd:element name="id" type="xsd:string"/>
<xsd:element name="system" type="xsd:string"/>
</xsd:sequence>
</xsd:complexType>
</xsd:element>
</xsd:sequence>
</xsd:complexType>
</xsd:element>

<!--ReliesOn-->
<xsd:element name="ReliesOn">
<xsd:complexType>
<xsd:sequence>
<xsd:element name="TrustNotion">
<xsd:complexType>
<xsd:sequence>
<xsd:element name="id" type="xsd:string"/>
</xsd:sequence>
</xsd:complexType>
</xsd:element>
<xsd:element name="Mechanism" minOccurs="0">
<xsd:complexType>
<xsd:sequence>
<xsd:element name="id" type="xsd:string"/>
<xsd:element name="system" type="xsd:string"/>
</xsd:sequence>
</xsd:complexType>
</xsd:element>
</xsd:sequence>
</xsd:complexType>
</xsd:element>
```

Fig. 6. XML schema of causal graph data model.

```
<!—Uses—>
<xsd:element name="Uses">
<xsd:complexType>
<xsd:sequence>
<xsd:element name="Mechanism">
<xsd:complexType>
<xsd:sequence>
<xsd:element name="id" type="xsd:string"/>
<xsd:element name="system" type="xsd:string"/>
</xsd:sequence>
</xsd:complexType>
</xsd:element>
<xsd:element name="Configuration" maxOccurs="1">
<xsd:complexType>
<xsd:sequence>
<xsd:element name="id" type="xsd:string"/>
<xsd:element name="system" type="xsd:string"/>
</xsd:sequence>
</xsd:complexType>
</xsd:element>
</xsd:sequence>
</xsd:complexType>
</xsd:element>

<!—DerivesFrom—>
<xsd:element name="DerivesFrom">
<xsd:complexType>
<xsd:sequence>
<xsd:element name="Capability">
<xsd:complexType>
<xsd:sequence>
<xsd:element name="id" type="xsd:string"/>
</xsd:sequence>
</xsd:complexType>
</xsd:element>
<xsd:element name="Mechanism" minOccurs="0">
<xsd:complexType>
<xsd:sequence>
<xsd:element name="id" type="xsd:string"/>
<xsd:element name="system" type="xsd:string"/>
</xsd:sequence>
</xsd:complexType>
</xsd:element>
</xsd:sequence>
</xsd:complexType>
</xsd:element>

</xsd:choice>
</xsd:complexType>
</xsd:element>
</xsd:schema>
```

Fig. 6. (*continued*)

```xml
<causal_graph_data>

<ReliesOn>
<TrustNotion> <id>confidentiality</id> </TrustNotion>
<Mechanism> <id>cpe:/a:example:disklocker:1.0</id>
<system>PHD_MC355_004</system> </Mechanism>
</ReliesOn>

<DerivesFrom>
<Capability> <id>disk_encryption</id> </Capability>
<Mechanism> <id>cpe:/a:example:disklocker:1.0</id>
<system>PHD_MC355_004</system> </Mechanism>
</DerivesFrom>

<CallsOn>
<MainMechanism> <id>cpe:/a:example:disklocker:1.0</id>
<system>PHD_MC355_004</system> </MainMechanism>
<SubMechanism> <id>cpe:/a:tpm:subsystem_key_generation:2.0</id>
<system>PHD_MC355_004</system> </SubMechanism>
<SubMechanism> <id>cpe:/a:tpm:subsystem_nv_memory:2.0</id>
<system>PHD_MC355_004</system> </SubMechanism>
<SubMechanism> <id>cpe:/a:tpm:subsystem_rng:2.0</id>
<system>PHD_MC355_004</system> </SubMechanism>
<SubMechanism> <id>cpe:/a:tpm:subsystem_symmetric_engine:2.0</id>
<system>PHD_MC355_004</system> </SubMechanism>
</CallsOn>
<Uses>
<Mechanism> <id>cpe:/a:example:disklocker:1.0</id>
<system>PHD_MC355_004</system> </Mechanism>
<Configuration> <id>CCE-071015-1</id>
<system>PHD_MC355_004</system> </Configuration>
</Uses>
<Uses>
<Mechanism> <id>cpe:/a:tpm:subsystem_key_generation:2.0</id>
<system>PHD_MC355_004</system> </Mechanism>
<Configuration> <id>CCE-071015-2</id>
<system>PHD_MC355_004</system> </Configuration>
</Uses>
<Uses>
<Mechanism> <id>cpe:/a:tpm:subsystem_nv_memory:2.0</id>
<system>PHD_MC355_004</system> </Mechanism>
<Configuration> <id>CCE-071015-3</id>
<system>PHD_MC355_004</system> </Configuration>
</Uses>
<Uses>
<Mechanism> <id>cpe:/a:tpm:subsystem_rng:2.0</id>
<system>PHD_MC355_004</system> </Mechanism>
<Configuration> <id>CCE-071015-4</id>
<system>PHD_MC355_004</system> </Configuration>
</Uses>
<Uses>
<Mechanism> <id>cpe:/a:tpm:subsystem_symmetric_engine:2.0</id>
<system>PHD_MC355_004</system> </Mechanism>
<Configuration> <id>CCE-071015-5</id>
<system>PHD_MC355_004</system> </Configuration>
</Uses>
</causal_graph_data>
```

Fig. 7. XML representation of the sample causal graph.

1. Is Capability id="Disk Encryption" ∈ of a triplet containing (TrustNotion="Confidentiality", RO)
2. Is Mechanism id="cpe:/a:example:disklocker:1.0" system="PHD_MC355_004" ∈ of a triplet containing (Capability id="Disk Encryption", DF)
3. Is Configuration id="CCE-071015-1" system="PHD_MC355_004" ∈ of a triplet containing (Mechanism id="cpe:/a:example:disklocker:1.0" system="PHD_MC355_004",U)
4. Is SubMechanism id="cpe:/a:tpm:subsystem_key_generation:2.0" system="PHD_MC355_004" ∈ of a triplet containing (MainMechanism id="cpe:/a:example:disklocker:1.0" system="PHD_MC355_004", CO)
5. Is Configuration id="CCE-071015-2" system="PHD_MC355_004" ∈ of a triplet containing (Mechanism id="cpe:/a:tpm:subsystem_key_generation:2.0" system="PHD_MC355_004", U)
6. Is SubMechanism id="cpe:/a:tpm:subsystem_nv_memory:2.0" system="PHD_MC355_004" ∈ of a triplet containing (MainMechanism id="cpe:/a:example:disklocker:1.0" system="PHD_MC355_004", CO)
7. Is Configuration id="CCE-071015-3" system="PHD_MC355_004" ∈ of a triplet containing (Mechanism id="cpe:/a:tpm:subsystem_nv_memory:2.0" system="PHD_MC355_004", U)
8. Is SubMechanism id="cpe:/a:tpm:subsystem_rng:2.0" system="PHD_MC355_004" ∈ of a triplet containing (MainMechanism id="cpe:/a:example:disklocker:1.0" system="PHD_MC355_004", CO)
9. Is Configuration id="CCE-071015-4" system="PHD_MC355_004" ∈ of a triplet containing (Mechanism id="cpe:/a:tpm:subsystem_rng:2.0" system="PHD_MC355_004", U)
10. Is SubMechanism id="cpe:/a:tpm:subsystem_symmetric_engine:2.0" system="PHD_MC355_004" ∈ of a triplet containing (MainMechanism id="cpe:/a:example:disklocker:1.0" system="PHD_MC355_004", CO)
11. Is Configuration id="CCE-071015-5" system="PHD_MC355_004" ∈ of a triplet containing (Mechanism id="cpe:/a:tpm:subsystem_symmetric_engine:2.0" system="PHD_MC355_004", U)

Fig. 8. Example of a trust policy.

```
for $c in
/causal_graph_data
where
$c/CallsOn/MainMechanism/id="cpe:/a:example:disklocker:1.0"
return
if ($c/CallsOn/SubMechanism/id="cpe:/a:tpm:subsystem_key_generation:2.0") then
<result1> yes </result1>
else
<result1> no </result1>
```

Fig. 9. XQuery command to check for a specific mechanisms that "DiskLocker" calls on.

References

1. National Cyber Leap Year Summit 2009 Report. https://www.nitrd.gov/
2. Grawrock, D., Vishik, C., Rajan, A., Ramming, C., Walker, J.: Defining trust evidence: research directions. In: Proceedings of the Seventh Annual Workshop on Cyber Security and Information Intelligence Research, p. 66, ACM (2011)
3. Stoneburner, G.: Underlying Technical Models for Information Technology Security. Recommendations of the National Institute of Standards and Technology, December 2001
4. Pearl, J.: Causality. Cambridge University Press, New York (2009)
5. Lewis, D.: Causation. J. Philos. **70**(17), 556–567 (1973)
6. Halpern, J.Y., Pearl, J.: Causes and explanations: a structural-model approach. part I: causes. Br. J. Philos. Sci. **56**(4), 843–887 (2005)
7. Trusted Computing Group. TNC Architecture for Interoperability. Specification Version 1.5. Revision 4, 7 May 2012
8. Trusted Computing Group. TNC IF-MAP Metadata for Network Security. Specification Version 1.1. Revision 9, 7 May 2012
9. Trusted Computing Group. IF-MAP Metadata for ICS Security. Specification Version 1.0. Revision 46, 15 September 2014
10. Will, A., Challener, D.: A Practical Guide to TPM 2.0: Using the Trusted Platform Module in the New Age of Security. Apress, New York (2015)
11. National Institute of Standards and Technology. National Vulnerability Database. Common Platform Enumeration. https://nvd.nist.gov/cpe.cfm
12. National Institute of Standards and Technology. National Vulnerability Database. Common Configuration Enumeration. https://nvd.nist.gov/cce/
13. University of Applied Sciences and Arts, Hochschule Hannover. Faculty IV, Department of Computer Science. http://trust.f4.hs-hannover.de/
14. Kim, A., Luo, J., Kang, M.: Security ontology for annotating resources. In: Meersman, R. (ed.) OTM 2005. LNCS, vol. 3761, pp. 1483–1499. Springer, Heidelberg (2005)
15. Yap, J.Y., Tomlinson, A.: A socio-technical study on user centered trust notions and their correlation to stake in practical information technology scenarios. In: Proceedings of the 6th ASE International Conference on Privacy, Security and Trust, 14–16 December 2014
16. Rivest, R.L.: Learning decision lists. Mach. Learn. **2**(3), 229–246 (1987)

Trusted Technologies

An Application-Oriented Efficient Encapsulation System for Trusted Software Development

Zheng Tao[1,2,3], Jun Hu[1,2,3]([✉]), Jing Zhan[1,2,3], Mo Li[1,2,3], and Chunzi Chen[1,2,3]

[1] College of Computer Science, Beijing University of Technology,
Beijing 100124, China
[2] Beijing Key Laboratory of Trusted Computing, Beijing 100124, China
[3] National Engineering Laboratory for Critical Technologies of Information Security
Classified Protection, Beijing 100124, China
algorist@bjut.edu.cn

Abstract. Trusted computing provides an efficient and practical way out for system security problems based on a trusted hardware, namely the root of trust, e.g., Trusted Platform Module (TPM), Trusted Cryptographic Module (TCM), Trusted Platform Control Module (TPCM), so on and so forth. However, current applications calling for trusted functions have to use either the user-space trusted interfaces (e.g., Trusted Software Stack (TSS) API) or to implement customized APIs on top of the trusted hardware driver; both of them are well known of steep learning curve, which indicates error prone and low-efficient development and complex maintenance for the application of trusted software. This paper presents a new trusted encapsulation architecture and the proof-of-concept system with the aim to mitigate the gap between the current obscure trusted APIs and the actual trusted applications for trusted software development. Our system can provide high-level and much simplified trusted transaction interfaces for user applications, which can rapidly reduce the development and maintenance work for the developers and users without too much performance costs. We also present a secure remote login use-case using mainly the binding and unbinding trusted functions of our trusted encapsulation architecture.

Keywords: Trusted computing · Application-oriented · Trusted software development

1 Background

Information security is a matter of national security and social stability. How to keep the information system from malicious attacks becomes an urgent problem to be solved. Trusted computing presents an on-going developing security methodology for system security [1–3]. The Trusted Computing Group (TCG)s Definition of "trust" is that an entity can be trusted if it always behaves in the expected manner for the intended purpose. In order to determine if a system is trusted, all of its components should work separately and together

© Springer International Publishing Switzerland 2016
M. Yung et al. (Eds.): INTRUST 2015, LNCS 9565, pp. 153–168, 2016.
DOI: 10.1007/978-3-319-31550-8_10

as expected. The common way to do this is by adding a trusted subsystem to the original system to make sure all the system components are not tampered from the initial clean slate. Normally, the trusted subsystem includes a root of trust which is a physically planted hardware, and corresponding software calling for the trusted interfaces of the hardware to extend the root of trust to the whole system through chain transition by verifying every components integrity when it launches [4,5] which is pretty simple idea but also a practical one. Furthermore, software developers can also call for the trusted interfaces of hardware to achieve application security with more assurance, since (1) the system integrity is protected and (2) sensitive application operations (e.g., encrypted storage of keys, decryption of small data) only occur inside the trusted hardware, even system administrators privilege cant subvert them.

Although trusted computing provides an efficient and practical way out for system and application security problems, and TCG/other stakeholders provide specific specifications [5] on how to develop with trusted APIs, trusted application is very rare on the market. Thiss mainly because of the complexity of the trusted software development, which involves different trusted hardware, e.g., Trusted Platform Module (TPM) [7,9], Trusted Cryptographic Module (TCM) [10], and Trusted Platform Control Module (TPCM), so on and so forth. Furthermore, current applications calling for trusted functions have to use either the user-space trusted interfaces (e.g., TCG Software Stack (TSS) API, TSM API [10]) or to implement customized APIs on top of the trusted hardware driver, both of them are well known of steep learning curve, which indicates error prone and low-efficient development and complex maintenance for the application of trusted software. Our paper focuses on how to develop the trusted software with high efficiency using trusted computing features.

Specifically speaking, when applications use a function of trusted computing, trusted interface calls trusted components to access the trusted root and other trusted computing resources. Trusted components exist as library functions or service processes. The typical trusted interface are provided to the upper calls, such as TCG using the trusted software stack (TSS), and China use the TCM service module (TSM), is directly facing the internal objects of the trusted root such as key object, PCR object, policy object and so on. A specific trusted computing function, usually requires multiple platforms, a variety of the trusted objects, and follows the specific protocols. It can be completed by some old trusted messages and external input messages. Therefore, in order to implement the function of trusted computing with trusted interface based on the standards like TSS, applications must manage various trusted objects and their associated relationships, and design some trusted protocols. Therefore, the implementation process inevitably involves a large number of trusted protocols and the details of the trusted object. So the realization of the trusted functions is extremely complex and it leads to the difficulties about development and maintenance of trusted computing [6].

Contribution. Based on five layers encapsulation architecture, this paper presents an application-oriented encapsulation system, which can provide very simple application-oriented interfaces for the user applications. With this system, applications can use trusted computing function expediently without get to the details of low-level trivial trusted APIs and we hope it can give a boost to real-world trusted computing application.

Outline. In this paper, Sect. 2 discusses the related research, Sect. 3 analyses the requirements and design for the encapsulation system, then describes the details of the five layers encapsulation, Sect. 4 designs a secure remote login usage to show the process of this encapsulation system, Sect. 5 gives an evaluation for the workload and performance and Sect. 6 concludes this work with a summary.

2   Related Work

The TPM main Specification 1.2 [5] and the TSS Specification 1.2 [7] introduce the basic principle of TSS design. In the case of TSPI, TSS service provider (TSP) provides an interface (TSPI), makes application software can be accessed to a group TCG functions through the TSP. All TSPI functions with one or more handle object parameters locate a particular category instance. It can call a group of sets or obtain attributes methods to access to attributes. The TSPI defines the following classes: context class, policy class, TPM class, key class, encrypted data class (sealed or bound data), PCR composite class, NV RAM class and hash class [7]. Objects which are scheduled by TSPI interface are almost all entities of the TPM. In the TCG Architecture TPM Library Specification 2.0 [9], TPM 2 has support for additional cryptographic algorithms, enhancements to the availability of the TPM to applications, enhanced authorization systems, simplified TPM management and additional capabilities to enhance the security of platform services. But the new TSS has not published yet, so we could not do some practical trusted works. With the old TSS, calling process of TPM 2 to finish trusted computing functions is also complex. So in this paper, the implement is still according to TPM 1.2.

As a matter of fact, most interfaces provide by TSPI and TSM are set for the internal entity objects of the trusted root. As describes in Sect. 1, it's very difficult and complicated to fulfill a single purpose of trusted APIs based job. In the case of the bind and unbind process with TSS, we assume that there are two interactive nodes and the two nodes implement the function of bind and unbind in a distributed system. First of all, the two nodes need to complete the key exchange, and it relies on the generation of AIK. In the generation process for AIK, there need to create TSS context, connection and obtain TPM object, load the root key, set the SRK password and so on. It includes more than 700 lines of code. And this is just the generation process of AIK. These complicated sets and operations need for manual operation, these complex works brought difficulties for the development and maintenance of trusted computing in TSPI.

The work of this paper is similar to that of a simplified trusted software stack names TSS [6]. But TSS simplifies the TSS, this paper present a fully encapsulation beyond the TSS. Wenchang Shi defines a TSB [8] as the totality of trust support systems for system software on a computing platform, but has not described the trusted application development in depth.

3   Design Philosophy of Trusted Encapsulation System

In this section, we present an application-oriented efficient encapsulation architecture and system.

3.1   Requirements and Principle

To use the Trusted Computing functions, well describe the requirements and design principles for serving these requirements. This system requires the following requirements:

- Compatibility: this system should satisfy the common standards like TSS.
- Usability: this system should easy to understand and use.
- Maintainability: the code of this system should easy to maintain.

In this paper, by providing a simple application-oriented trusted interface, applications can call trusted functions easily. The trusted interface can calls the standard trusted root interface like TSS to access the TPM, TCM and other trusted root through five layers encapsulation. Thats to say, the principle of this systems design is to provide a simplified application-oriented trusted interface, and establishes connection between the interfaces and the local objects through five layers encapsulation to finish trusted computing processes.

In general, a trusted transaction implement through interactive process between multiple trusted components of different machines. Each trusted components manages trusted local objects and execute local trusted process. Meantime, these objects managed by trusted components are cryptography objects, such as input and output data object, key object, Hash object, etc. There is a connection between these objects in the sense of cryptography and they are implements through data entity objects (including the internal objects of trusted root and external data objects of the trusted root) associated with trusted root. In order to implement the interactive process of multi nodes and elements, the local objects management system and the trusted operation, cryptography objects and operation, mapping between trusted root entity objects and operation, and other functions, this system should implement a multi layers encapsulation.

3.2   Architecture

Based on the previous analysis, this paper designs a five layers encapsulation architecture, as shown in Fig. 1. The five layers are hardware encapsulation, interface encapsulation, plugin encapsulation, process encapsulation and application encapsulation, and it has implement the transparent trusted function interface in application layer. As shown in Fig. 1, the top layer provides an interface which can be called easily for applications; Under this layer, there includes the interactive process between different nodes; Under the interactive process, we designs a encapsulation for each trusted computing process for the concrete works in a single node and storage the relationships of different local objects in a database; In the next layer, we will encapsulate all kinds of work objects to TESI (Trust Encapsulated Service Interface, namely the encapsulated function interfaces) library; To the bottom software stack, the most important principle is to keep the compatibility with existing standards.

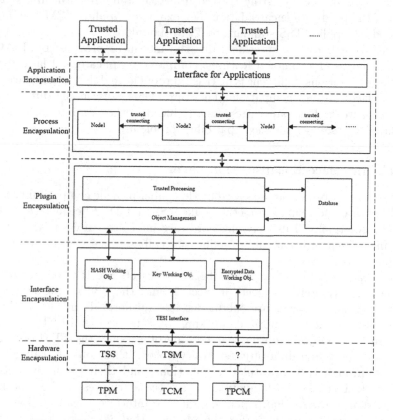

Fig. 1. Five-layer trusted computing encapsulation structure.

In this encapsulation system, we define the APIs which designs for TPM functions like TSS API and TSM API as hardware encapsulation system; interface encapsulation simplify the calling process of the trusted root through the classification of TSS work objects and encapsulation for the local library functions based on the hardware encapsulation; plugin encapsulation encapsulates the calling of local trusted computing functions and the management of trusted data in the manner to plugins; process encapsulation system implements the interaction between different nodes through trusted protocols by message policy configuration; application encapsulation associates the trusted requirements of applications with trusted computing process. Through the encapsulation system, trusted computing process can be triggered automatically and transparently.

3.3 The Hardware Encapsulation Layer

Hardware encapsulation realizes an interface which makes trusted roots functions can be accessed directly by application layer. The design based on compatible with existing standards, existing trusted interface standards still in use as the method of the hardware encapsulation. For example, Applies TCM standard to TCM module, applies TSS standard to TPM module.

But existing hardware interfaces like the interfaces provides by TSM and TSS, they have some common features, complex, complicate and lack of effective debug promotions. Different type trusted roots have different interfaces, and these interfaces are not compatible. These problems cause many difficulties on development work. This paper presents a reduced method on interface encapsulation layer to resolving the incompatibility problem.

3.4 The Interface Encapsulation Layer

The interface encapsulation presented by this paper provide compatibility for different hardware encapsulation methods like TSS and TSM because of the hardware encapsulation still use existing hardware interfaces. It hides the complex underlying implementation, and simplifies the calling process of the applications from upper layer.

In interface encapsulation layer, the local objects are the top priority. We described the division of objects in Sect. 2, and the division is designed by TSS. In TSS, local objects are subdivides into authorized and non-authorized working objects, non-authorized working objects are the PCR composite objects and hash objects, authorized working objects are the TPM object, key objects and encrypted data objects. In cryptology, these objects call standard interfaces through the encapsulation functions, and then reach the entity of trusted root, TSS use TSPI functions to set the attributes to the objects like Tspi_SetAttribData. In this papers design, the interface encapsulation encapsulates these setting methods for the objects, that is, when use an object, it is not necessary to set complex settings. For example, three TSPI functions, Tspi_Context_CreateObject, Tspi_Data_Bind and Tspi_GetAttribData are encapsulate in a TESI function, TESI_Local_Bind. In a binding process, by the origin called procedure of TSPI, the three functions used repetitively, and require 13 parameters, 70 source code lines. In this papers design, only require 3 parameters, and 70 source code lines compressed into a function.

The encapsulate process is as below:

1. According to application needs, the default binding and one-to-one correspondence parameter objects will encapsulate as internal variables. Then design the encapsulation function interfaces;

2. Encapsulate TSS/TSM or other trusted roots trusted software stack interface functions in the interface functions to realizes the interface;

3. To realize the interface function encapsulation library, at first write the interface functions to head files, then compile the code of interface functions to static library.

According to interface encapsulation and using the rules above, in the interface encapsulation layer, this system realizes a series of functions to support trusted functions such as AIKs generation, sign a file, verify a file, binding and unbinding, TSS/TSM or other trusted roots trusted software stack interface functions are encapsulate in the function interfaces. The complexity of interfaces is rapidly reduced, functions are more directly and clearly, and conceal the difference from different hardware encapsulations such as TSS and TSM, meet the design requirement.

As a consequence of the above, in the encapsulation of interface, every interface function encapsulates multiple TSPI functions. The numbers of calls in calling process are markedly reduced. Meanwhile, the facts mentioned above, many complex parameters and definitions are hidden, especially TSS exclusive definitions are not shown in the interface layer, parameters such as hContext and hsrk are recessively called as default parameters. The frequency of calling is decreased to 1/6.The result of the test showed that the interface encapsulation can rapidly reduce the complexity of TSS effectively, and hide the features of TSS.

3.5   The Plugin Encapsulation Layer

Trusted computing functions need to generate and process multiple local objects associate with trusted computing functions, such as different key values, PCR values and encrypted data and decrypted data. In Sect. 3.4, we have described the subdivision of local objects in TSS. But in our daily development, were only concerned about how to call the concrete objects like EK, AIK, its not just one key object. So in this papers design, we separate the key object, and the relationship is depicted as Fig. 2 shown.

In real applications, we need to call different objects to finish a work. To call the different objects, we should find the relationships between them. To implement the finding and management process, we design a database to store the local objects association. For example, the private and public keys storage structure is shown in Tables 1 and 2. From Tables 1 and 2, if a plugin need to find a keys corresponding virtual TPM, it can just querying the storage to find the virtual TPMs uuid to get the right virtual TPM. If this plugin reads a private key and want to get the corresponding public key, it can querying the public keys storage structure to find the public key with the same virtual TPM uuid.

The various local actions in trusted measurement, trusted storage and trusted report can encapsulate in a plugin, the database is used to store the association among trusted objects generation and management. This design offers standard interface functions with the predefined message and data format. As above, complex trusted processes are simplified, now the processes use message to driving

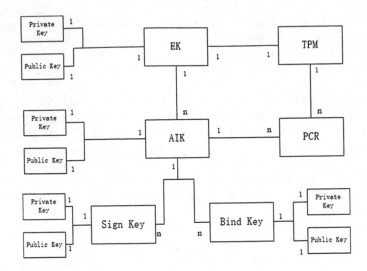

Fig. 2. Local object relationship.

Table 1. The private keys storage structure.

Data name	Type	Length	Explanation
uuid	char *	DIGEST_SIZE*2	Key identifier
vtpm_uuid	char *	DIGEST_SIZE*2	Associated virtual TPM identifier
issrkwrapped	bool	4	Is the key encrypted by SRK
key_type	int	4	Key type, identity key, sign key, binding key, etc.
key_alg	int	4	Password type, RSA or SM2
key_size	int	4	Key size
key_binding_policy_uuid	char *	DIGEST_SIZE*2	Bind policy identifier with key
wrapkey_uuid	char *	DIGEST_SIZE*2	Wrapkey identifier
keypass	char *	Unknown	Password
key_filename	char *	Unknown	File name of the key file

a series of plugin to work. Meanwhile, developers with different requirements can customize their own trusted plugin, and cooperate with other developers plugin to finish a work.

According to the above research frame, This paper designed a plugin encapsulation structure as Fig. 3.

As above, to implement the plugin encapsulation, the relationships among local objects need to store into the local database. The local objects management and local works processing will be encapsulate in a plugin which have input and output functions with message, and designed a local database for the plugin, the database protect its internal data storage by trusted storage technology. In this design, relationships among different local objects can be finding by

Table 2. The public keys storage structure.

Data name	Type	Length	Explanation
uuid	char *	DIGEST_SIZE*2	Key identifier
vtpm_uuid	char *	DIGEST_SIZE*2	Associated virtual TPM identifier
ispubek	bool	4	Is the key public key of EK
key_type	int	4	Key type, identity key, sign key, binding key, etc
key_alg	int	4	Password type, RSA or SM2
key_size	int	4	Key size
key_binding_policy_uuid	char *	DIGEST_SIZE*2	Bind policy identifier with key
privatekey_uuid	char *	DIGEST_SIZE*2	Corresponding private key identifier
keypass	char *	Unknown	Password
key_filename	char *	Unknown	File name of the key file

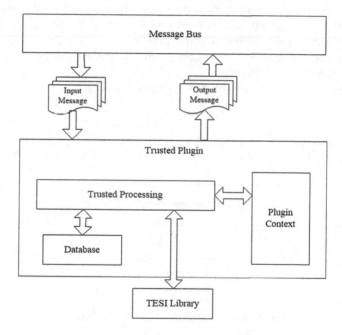

Fig. 3. Plugin encapsulation structure.

querying the flags defined in the data structure of the local objects. The process of querying the association among local objects from the local database and do some processing shows the substance of plugin encapsulation. The design of the database defines the storage structure for these objects.

3.6   The Process Encapsulation Layer

Plugin package encapsulates all local trusted processes, process encapsulation complete the interaction and collaboration of these plugins.

Process encapsulation is based on a message bus which has message routing function, this message bus can confirm the messages destination through the content, type, and the sender of the message, and it also can realize the message routing like challenge - response patterns, and the aspect-oriented message routing which can intercept the message from the existing message process and return this message after processing. The message bus can mount many trusted encapsulated plugins, and connect the plugins through the message routing, finally implement the trusted collaborative process automatically.

In this process, it is necessary to set policies for the message bus to deliver the message correctly. Message policies are divided into two patterns, match policy and route policy.

When the message bus receives a message, the match policy start to work. It compare the message with the pre-defined format, and classify the messages, then give the messages to different route policy.

Route policy confirms the messages from the match policy, then send the messages to the right destination.

The design of process encapsulation is shown in Fig. 4.

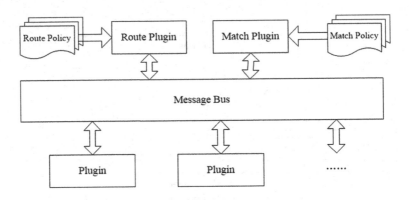

Fig. 4. Workflow of process encapsulation.

In this design, the trusted cooperation between different nodes can be encapsulated as an individual process through the interaction between different plugins by route policy defines, to make different plugins to complete a trusted process synergistically. That does not only simplify the development, but also makes the trusted encapsulation system runs automatically as software defined.

3.7   The Application Encapsulation Layer

On the basis of the preceding discussion, we can get a complete process of trusted computing functions. So in the application layer, we need to design an interface for applications to call the process conveniently. Application encapsulation is to translate the trusted requirements of the applications to local plugins that will implement these requirements and processes between those plugins. First, we need to implement the corresponding plugins, and mounts the plugins on the message bus, and then map the application demand to the corresponding process. In practice, such process was triggered when applications propose their demands to accomplish trusted requirements. Meantime, environmental factors and other security mechanism can also trigger other trusted process through application layer monitor service, thus making the trust functions of the application can reflect security changes of the environment and the effect of security system and security events have on it. Finally, based on plugin encapsulation, provide transparent support of trusted binding functions for applications through an application library.

4   Secure Remote Login Usage Example

This paper presents a five layers encapsulation system based on Linux environment and TCG standard, and designs a usage. The usage uses swTPM of IBM to simulates physical trusted root (TPM), and uses trousers to implement the TSS trusted software stack as physical layer encapsulation. Take the case of a process of remote login, the structure of this login system is shown in Fig. 5. The login node receives login message form user, and send this message to verify node. Verify node authenticate the user, and sends the result to the login node. For security reasons, we use bind method of trusted computing to realize the protection of data. Verify node generates a bind key (a cryptographic key pair, private key stores in the verify nodes trusted root and public key public), login node gets the public key of verify nodes bind key, and uses this public key to encrypts login message, then sends the encrypted message to verify node, verify node uses private key to decrypt login message. We can make sure that the login message can be only reads by the verify node because only verify nodes trusted root has this private key of bind key. For these reason, we can prevent any information from leaking out.

In the process of bind and unbind, there are 11 TESI functions was called. These functions are shown in Table 3.

Then we designed 5 plugins to finish the local processes: login plugin, AIK Client plugin, sign/verify Plugin, binding/unbinding plugin and authentication plugin. These plugins have deployed on different nodes. Every node has deployed some message buses. Plugins transport messages by message bus though the match policy and route policy. The deployment of different nodes is shown in Table 4.

The analysis of information interaction process between nodes is as follows:

Step 1. The process of negotiation certificate between the login node and authentication node.

1.1 Binding/unbinding node sends AIK application package containing certificate information to AIK authentication node;

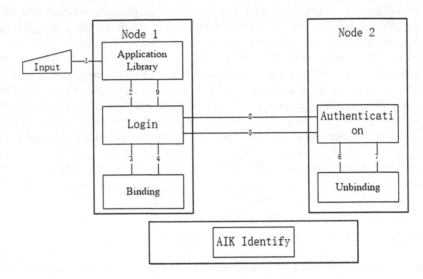

Fig. 5. Remote login system structure.

Table 3. TESI functions uses in bind/unbind.

Function name	Function
TSS_RESULT TESI_Local_Auth(char * pwdo, char * pwds)	Connect with trusted root, and get the permission of trusted root and storage root
TSS_RESULT TESI_Local_Reload()	Connect with trusted encapsulation library
TSS_RESULT TESI_Local_CreateBindKey(TSS_HKEY *hKey, TSS_HKEY hWrapKey, char * pwdw,Char * pwdk)	Create a bind key
TSS_RESULT TESI_Local_WriteKeyBlob(TSS_HKEY hKey, Char * name)	Write the private key to a file
TSS_RESULT TESI_Local_ReadKeyBlob(TSS_HKEY *hKey, Char * name)	Read the private key from a file
TSS_RESULT TESI_Local_WritePubKey(TSS_HKEY hKey, Char * name)	Write the public key to a file
TSS_RESULT TESI_Local_ReadPubKey(TSS_HKEY *hKey, Char * name)	Read the public key from a file
TSS_RESULT TESI_Local_LoadKey(TSS_HKEY *hKey, TSS_HKEY hWrapKey, Char * pwdk)	Use wrap key to unwrap the public key from a file
TSS_RESULT TESI_Local_Bind(char * plainname, TSS_HKEY *hKey, char * ciphername)	Bind data from a unencrypted file
TSS_RESULT TESI_Local_UnBind(char * plainname, TSS_HKEY *hKey, char * ciphername)	Unbind data from a encrypted file

Table 4. Plugins on different nodes.

Node name	Bus name of the node	Function of the bus	Plugins of the bus	Notes
Login node	Login bus	Accept login information and send to Verify Node	Login plugin	Accept and generate login message
	Trusted processing bus	AIK certificates request, active, sign/verify and binding/unbinding	AIK client plugin sign/verify plugin binding/unbinding plugin	
Verify node	Verify bus	Authentication	Authentication plugin	Verify login message and output the result as a verify message
	Trusted processing bus	AIK certificates request, active, sign/verify and binding/unbinding	AIK client plugin sign/verify plugin binding/unbinding plugin	
CA node	CA verify bus	AIK identify	AIK identify plugin	

1.2 AIK authentication node provides the AIK activation package containing the signature certificate to binding/unbinding node;

1.3 Binding/unbinding nodes activates the authentication key and get the signature certificate.

Step 2. The process of keys exchange between binding node and unbinding node.

2.1 Unbinding node sends the certificate containing signature public key to binding node;

2.2 Binding node extracts the signature public key from the certificate after completes the authentication of certificate message;

2.3 Unbinding node generates a pair of binding key, and then sign and issue the public key of binding key with signature key, at last sends it to binding node;

2.4 Encryption binding node obtains binding public key and verifies it with signature.

Step 3. The process of login node send login messages, and verification node authenticate and return the result.

3.1 The login node obtains user's login information from external input;

3.2 Login node sends message to binding node;

3.3 Binding node sends the message data to login node after encrypting it;

3.4 Login node sends encrypted message to verification node;

3.5 Verification node sends the encrypted message to unbinding node, and unbinding node decrypts the message;

3.6 Unbinding node sends the decrypted data to verification node;

3.7 Verification node generates the return message through verifies the login behaviour according to the message, and sends the message to login node;

3.8 Login node reads the return message and finishes the process of verifying.

As the process described above, by using an interface from application layer, an application can just input a user name and a password, then wait for the reporting of verification.

The two examples below shows how to bind and unbind some data through this encapsulation system.

Example of a binding data process

```
try {
cout <<"TESI_bind_binddata" << endl ;
result=TESI_Local_Reload();
}
catch (!TSS_SUCCESS &e ) {
cerr << "Tspi_Key_LoadKey:" << e << endl ;
}
try {
result= TESI_Local_ReadPubKey(&hKey,"bindpubkey");
}
catch (!TSS_SUCCESS &e ) {
cerr << "Tspi_Key_LoadKey:" << e << endl ;
}
try {
cout << "Data before binding" << endl ;
result=TESI_Local_Bind("plain.txt",hKey,"cipher.txt");
}
catch (!TSS_SUCCESS &e ) {
cerr << "TESI_Local_Bind:" << e << endl ;
}
```

Example of an unbinding data process

```
try {
cout << "TESI_bind_unbinddata" << endl ;
result= TESI_Local_ReloadWithAuth("a","b");
}
catch (!TSS_SUCCESS &e ) {
cerr << "Tspi_Key_LoadKey:" << e << endl ;
}
try {
result= TESI_Local_ReadKeyBlob(&hKey,"bindkey");
}
catch (!TSS_SUCCESS &e ) {
cerr << "Tspi_Key_LoadKey:" << e << endl ;
}
try {
result= TESI_Local_LoadKey(hKey,NULL,"k");
```

```
}
catch (!TSS_SUCCESS &e ) {
cerr << "Tspi_Key_LoadKey:" << e << endl ;
}
try {
result=TESI_Local_UnBind("cipher.txt",hKey,"plain.txt");
TESI_Local_Fin();
}
catch (!TSS_SUCCESS &e ) {
cerr << "TESI_Local_UnBind:" << e << endl ;
}
```

5 Evaluation

In this test, through survey of the workload by use TESI library or use TSPI library directly, we can concluded that use TESI library functions can reduce 2/3 code on average. And in the application layer, by using an application encapsulation interface to finish the same work, we need just a single command. Some statistical results are shown in Table 5.

Table 5. Workload in different calling method.

	No encapsulation	Using encapsulation	Application call
AIK process	1226 code lines	285 code lines	A single command
Binding key generate	202 code lines	60 code lines	
Binding/unbinding data	1068 code lines	399 code lines	

All these processes runs in the memory, so there are no significant differences between encapsulation and no encapsulation in performance.

6 Conclusion

This paper design a architecture and implement a proof-of-concept system with the aim to mitigate the gap between the current obscure trusted APIs and the actual trusted applications for trusted software development. This system can provide high-level and much simplified trusted transaction interfaces for user applications through five layers encapsulation. In the architecture, this system imposes four encapsulations: interface encapsulation, plugin encapsulation, process encapsulation and application encapsulation on the basis of the original trusted hardware encapsulation, and provides simplified application-oriented interfaces, realized a simplified way to call trusted computing functions, solved the compatibility problem among different trusted hardware encapsulation interfaces, hidden the details of local trusted computing process and keys management, implemented the automation of trusted protocol in different nodes and the translate process from application trusted requirements to trusted computing functions, finally realized transparent trusted computing support for applications.

In summary, with this encapsulation system, developers can use trusted computing functions expediently and efficiently. We hope the encapsulation architecture we proposed can give a boost to real-world trusted computing application.

Acknowledgments. This work is supported by grants from the China 863 Hightech Programme (Project No. 2015AA016002) and Specialized Research Fund for the Doctoral Program of Higher Education (Project No. 20131103120001).

References

1. Shen, C., Zhang, H., Wang, H., et al.: Study of trusted computing and its development. Sci. China Inf. Sci. **40**(2), 139–166 (2010). (in Chinese)
2. Zhang, K., et al.: Reconfigurable security protection system based on NetFPGA and embedded soft-core technology. In: The International Conference on Computer Design and Applications (ICCDA 2010), vol. 5, pp. 540–544 (2010)
3. Feng, D., Yu, Q., Wei, F., et al.: The theory and practice in the evolution of trusted computing. Sci. Bull. **59**(32), 4173–4189 (2014)
4. Berger, B.: Trusted computing group history. Inf. Secur. Tech. Rep. **10**, 59–62 (2005)
5. Trusted Computing Group: TPM main specification. Main specification version 1.2 rev. 103, Trusted Computing Group, July 2007
6. Stüble, C., Zaerin, A.: μTSS – a simplified trusted software stack. In: Acquisti, A., Smith, S.W., Sadeghi, A.-R. (eds.) TRUST 2010. LNCS, vol. 6101, pp. 124–140. Springer, Heidelberg (2010)
7. Trusted Computing Group: TCG Software Stack (TSS) specifiction, version 1.2, Errata A[EB/OL] (2009). http://www.trustedcomputinggroup.org/resources/tcg-SOftware_stack-tss_specification
8. Shi, W.: On design of a trusted software base with support of TPCM. In: Chen, L., Yung, M. (eds.) INTRUST 2009. LNCS, vol. 6163, pp. 1–15. Springer, Heidelberg (2010)
9. Trusted Computing Group: TCG architecture TPM library specification 2.0 (2014). http://www.trustedcomputinggroup.org/resources/tpm_library_specification/
10. State Cryptography Administration Office of Security Commercial Code Administration (OSCCA): Functionality and Interface Specification of Cryptographic Support Platform for Trusted Computing. http://www.oscca.gov.cn/UpFile/File64.PDF

Research on Trusted Bootstrap Based on the Universal Smart Card

Lin Yan[1,2,3](✉) and Jianbiao Zhang[1,2,3](✉)

[1] College of Computer Science, Beijing University of Technology, Beijing, China
chriszt@126.com, zjb@bjut.edu.cn
[2] Beijing Key Laboratory of Trusted Computing, Beijing, China
[3] National Engineering Laboratory for Critical Technologies of Information Security
Classified Protection, Beijing, China

Abstract. The trusted boot is a hot spot in trusted computing field. User's identity authentication and trusted measurement are used to deal with security threats. But it is difficult to implement the general trusted boot based on hardware, which can be bypassed easily by software. In order to solve the above problem, a scheme of trusted boot is presented based on the universal smart card. It does not change the hardware and the firmware of the smart card and the terminal device. The core method combines user's identity authentication with trusted measurement. It binds user's identity, smart card and terminal device to ensure the trusted boot of terminal device. The trusted computing mechanism can be extended from power on to the application layer. Ultimately, experiments prove the security of boot and simplification of the implementation.

Keywords: Trusted computing · Trusted root · Trusted chain · Trusted measurement · Security bootstrap

1 Introduction

With the development of information technology, information security problems have become increasingly serious [1,2]. Desktop and laptop computers that are most commonly used store critical data as the terminal device for users. Therefore, in order to solve security issues, it must be started from the terminal device. Due to the security mechanism of hardware structure of the terminal device is oversimplified and the terminal device lacks design for security [3,4]. The attacker can boot the terminal device without the permission, to bypass the authentication mechanism of the operating system, and steal or tamper with the user's critical data and damage operating environment in terminal device. In order to prevent above attacks fundamentally, it must protect the terminal device from power on to the application [5,6]. Unless protecting the integrity of the entire boot mechanisms of terminal device, it can protect the security of critical data effectively [7].

© Springer International Publishing Switzerland 2016
M. Yung et al. (Eds.): INTRUST 2015, LNCS 9565, pp. 169–182, 2016.
DOI: 10.1007/978-3-319-31550-8_11

In terms of trusted boot, many scholars and institutions at home and abroad have made a great contribution to the prosperity. Literature [8,9] presented the cryptographic operation function is introduced to the CPU, in order to make the CPU have the logical and cryptographic operations functions. However, the scheme needed to transform the existing hardware architecture of the CPU, and it did not have the generality. Literature [10,11] needed the help of a trusted third party, the introduction of a trusted server, and realized a trusted boot with the terminal via trusted server. However, the program needed trusted third party authentication, and trusted measurement was achieved for online and complex trusted systems was involved. Trusted Computing Group (TCG) proposed Trusted Computing Module (TPM) [12]. When the computer was power on, the core root trusted module (CRTM) measured the BIOS, and when the integrity of BIOS satisfied, the process passed to the BIOS. Then the BIOS measured the operating system loader (OS Loader), the OS Loader archived the control when the integrity of the Boot Loader meet. Ultimately, the system completed the transfer of trust chain. TCG proposed that the trusted boot mechanism had three forms on the hardware platform: Firstly, The TPM chip was embedded on the motherboard [13], and connected to the Southbridge chip via LPC (Low Pin Count) bus. However, the scheme needed to change the hardware structure and only to protect the security of the new machine, but was not compatible with the hardware environment of the old machine; secondly, External PCI card was connected to the motherboard [14], and it could protect the security of the old machine. But the scheme was only to protect the security of desktop computers, laptop computers could not be on an external PCI card, the security of laptop computers was not ensured; thirdly, the form of USB interface [15] could simultaneously protect the security of desktop and laptop computers. But the scheme needed to modify the BIOS program, and because of the manufacturer of BIOS were different, the transformation capacity and operability was not strong.

Based on the third scheme, Chinese scholars [16–19] presented a new scheme which did not change the hardware architecture and BIOS program combined the BIOS with the USB device together as the trusted root, and then established a trusted chain during startup terminal device. But the scheme had the following shortcomings: Firstly, the dedicated USB device must be customized and large overhead costs; secondly, there was a security risk in that multiple terminal devices could be boot by the same USB device; thirdly, if the boot data of terminal device had been tampered with or destroyed, then no recovery mechanism was taken into account; and lastly, the USB device was easy to lose and damage, without taking into account the recovery mode of the USB device itself.

This paper is based on the plan put forward by Chinese scholars which did not change the hardware architecture of the smart card. The smart card that includes CD-ROM file system transforms with storage file to design the trusted CD boot (TCDB) and the transformation of the partition boot sector (PBR). Without changing the hardware and firmware architecture of the smart card and the terminal device, the security objective is implemented. It binds user's identity information to the terminal device and the TCDB. The trusted mechanism could be extended from the power on of terminal device to the application layer. The solution is also presented for situations where PBR is tampered with and TCDB is lost or damaged. The paper describes the transformation scheme for TCDB logical structure and disk boot program, and then introduces a trusted boot mechanism of terminal device and the credibility of the mechanism is proved. Finally through the theoretical analysis and experimental data, the safety and practicality of the scheme are proved.

2   Logical Structure of TCDB

The traditional smart card with CD-ROM file system includes a storage module and a COS (Chip Operating System) module. As shown in Fig. 1.

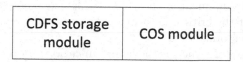

Fig. 1. Traditional smart card hardware logic structure

Without changing the hardware architecture of the smart card, the storage module includes the boot of the program and data, extends the capabilities of the smart card, after the transformation of the logical structure as shown in Fig. 2.

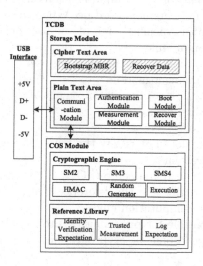

Fig. 2. Traditional smart card hardware logic structure

The storage module is responsible for taking over the boot process from the terminal device, verifying user's identity information and measuring the platform configuration information. It also provides a function that recovers platform configuration information. The storage module is logically divided into two parts: the cipher text storage area and the plain text storage area.

- The cipher text storage area

In order to prevent the attacker copying the critical data including the system boot MBR and the recovered data, the critical data are stored in the cipher text storage area.

1. The system boot MBR

The transformation of the MBR in disk is that it removes the instructions which include judging the device type and loading the PBR, in order to improve the boot speed. When the execution is completed, the control will jump to the specified area of memory.

2. The recovered data

If the trusted measurement of the terminal device is not satisfied, the recovered data will repair the critical data in the terminal device.

- The plain text storage area

All of the logic modules such as the communication module, the boot module, the authentication module, the trusted measurement module and the recovery module are stored in this area.

1. The communication module

This module calls COS module to complete the data encryption and decryption operations. Simultaneously, the module is responsible for the communication between TCDB and the terminal device.

2. The boot module

This module takes over the boot process of the terminal device, and provides the execution environment for the communications module, the authentication module and the trusted measurement module. When the execution is completed, the control will jump to the specified area of the memory.

3. The authentication module

This module is responsible for authenticating user's identity information.

4. The trusted measurement module

This module is responsible for the measurement of the platform configuration information of the terminal device.

5. The recovery module

If the terminal device does not satisfy the trusted measurement process, this module will repair critical data.

The COS module is a system on chip (SOC). This module includes the cryptographic computation engine and the reference library.

- The cryptographic computation engine

The engine is responsible for providing cryptographic operations, and supporting our self-developed domestic cryptographic algorithms.

• The reference library

The library includes expected values of authentication, trusted measurement and log information of the three parts. It provides the baseline for the user authentication and platform trusted measurement and it also provides the trusted proof for the application layer.

3   New Boot Sectors Structure

Typically, boot sectors can be run automatically when the terminal device powers on. The structure of boot sectors is shown in Fig. 3.

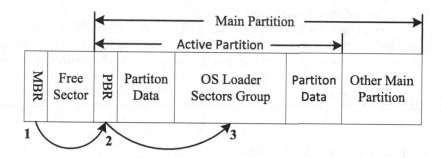

Fig. 3. Boot sectors of disk

When booting, the operating system first executes the MBR program to find the active partition; then executes PBR in the active partition, locates the operating system loader sectors group; and finally executes the OS Loader.

In order to prevent the terminal device from starting alone; it must revise the boot sectors. Revised sectors include MBR, PBR, and the OS Loader sectors group. After the transformation, the disk structure is shown in Fig. 4.

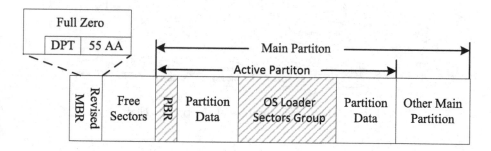

Fig. 4. Boot sectors after transformation

Both data section and code section of MBR are set to 0, saving the disk partition table (DPT) and the end identifier 0×55 and $0 \times AA$. When the terminal device starts alone, the boot process can be prevented.

If an attacker restores MBR in disk artificially, by encrypting PBR of the active partition, the boot process cannot continue. As shown in the shaded portion of PBR in Fig. 4. In this way, the security of some metadata such as the file system identification and the sector number of a cluster of disk can be protected. In addition, the read logic in the PBR sector is modified, in order to implement the execution process which can jump to the specified area of the memory.

If the MBR and the PBR of the active partition of disk have been restored artificially, the OS Loader sectors group will be encrypted to prevent the boot process from going on. As shown in the shaded portion of OS Loader sectors group in Fig. 4. OS Loader consists of the executable binary code and multiple configuration files, so for the attacker to restore the sectors is difficult. In this way, the terminal device cannot startup alone.

When the terminal device is started, the TCDB gets the booting process and measures the transformation of MBR, PBR, and the OSLoader sectors group in order to proof the security. If the result cannot meet the requirements of credibility, the user can choose whether to recover data in three regions of disk from the TCDB. In this way, the boot sectors of the terminal device can be protected.

4   Trusted Boot Mechanism

Binding user's identity, TCDB and terminal device is to ensure the security of the terminal device from starting up to and before loading the operating system kernel. The transferring process of the trusted chain is shown in Fig. 5.

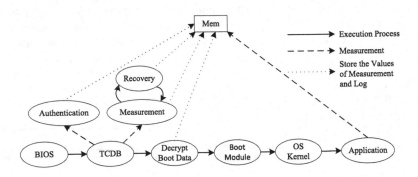

Fig. 5. The transfer process of the trusted chain

Without alteration to the BIOS, the trusted boot are divided into four stages which includes establishing the trusted root, the trusted measurement, the trusted recovery and the trusted execution.

1. Establishing the trusted root

When the BIOS load is completed, the binding TCDB gains the control. The user inputs the authentication information which is verified with the expected value stored in the binding TCDB. If the process is satisfied, the BIOS and TCDB will combine together as the trusted root of the terminal device. Otherwise, if the TCDB which should bind to the terminal device does not connect to the terminal device or the user authentication fails, the boot process of the terminal device will be terminated.

2. The trusted measurement

The TCDB measures the platform environmental information of the terminal device. If the step fails, the control will enter the trusted recovery stage. Otherwise, it will enter the trusted execution stage.

3. The trusted recovery

This takes into account the abnormal situation that the terminal device connects to the unbinding TCDB in order to enhance the robustness of the system. The trusted recovery is divided into two parts: the user verification and the platform recovery.

• The user verification

In order to prevent the TCDB from recovering the boot data in the disk caused by the user who inserts the unbinding TCDB to the terminal device, the user must consult with the trusted third party. When the trusted third party checks on the information whether the user submits correctly, the process of execution will pass to the platform recovery step.

• The platform recovery

One part of recovering data stored in the cipher text area is decrypted by the COS module to repair the MBR of the disk. Then the other parts of the recovering data repair the PBR and OS Loader sector group. When the recovery step is completed, the TCDB will measure the platform environmental information of the terminal device once again.

4. The trusted execution

The communication module in the TCDB decrypts the MBR stored in the cipher text area and PBR stored in the active partition of the disk, and then copies the plain text of them to the specified area of the memory. It also decrypts the OSLoader sector group and extracts the executable binary data from the plain text to recombine the OS Loader and copies it to the specified area of the memory as shown in Fig. 6.

When the boot module finishes, the MBR decrypted in the memory will execute, followed by PBR, the OS Loader and then load the OS kernel. Even if power is off in the process of bootstrap, the TCDB and boot sectors of the disk will still cipher text storage. In this way, the integrity of critical data can be protected.

After the operating system is booted, the application of user level must prove that the bootstrap is trusted before operation of the terminal device. This process includes two parts: the credibility of the measuring log and the TCDB.

• The credibility of the measuring log

In order to ensure the credibility of the bootstrap, the corresponding log information is stored in memory. Comparing the log in memory with the expected value in the TCDB can ensure the credibility of the bootstrap.

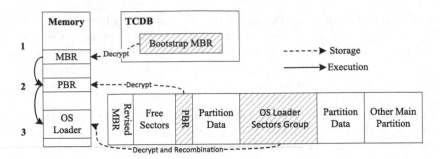

Fig. 6. Trusted execution process

• The credibility of the TCDB

Considering the situation that the TCDB is lost or damaged, the TCDB must be re-created. At present, the new TCDB does not match the terminal device. After the bootstrap of the operating system completed, the application layer must re-bind the relationship between user's identity, TCDB and terminal device. In this way, the lost or damaged TCDB cannot be reused.

5   Proof of the Credibility

In order to ensure credibility of the scheme, the proof of the credibility of the bootstrap is needed. Many research institutions have their own definition of trusted computing. This paper uses definition proposed by Chinese scholars. The core of building the trusted root is to ensure that the user's identity information meet the requirements of TCDB. The core of the trusted measurement is to ensure the platform configuration information of the terminal device meet TCDB requirements. Therefore, some basic concepts need to be defined: user's identity information and the platform configuration information.

According to the definition of Chinese scholars [20], the definitions of the trusted boot are the followings:

Definition 1: The trusted boot
If each step of the boot can meet the needs of the expected values, and the system can prove the result is credible, then the boot process is the trusted boot.

Definition 2: User's identity information
It is personal information that the user satisfies certain functional requirements, represented by such as PIN code. The set is constituted by user multiple identities information represented by $R = \{r_1, r_2, \cdots, r_i, \cdots, r_n\}$.

Definition 3: The platform configuration information
Using the quad $c_i = <s_i, e_i, d_i, h_i>$ describes the platform configuration information of terminal device, s_i represents the starting number of boot sectors in disk; and e_i represents the end number of boot sectors in disk, and $s_i \leq e_i$; and d_i represents the data of sectors group $[s_i, e_i]$; and h_i represents the hash value of

d_i; the set constructed by the platform configuration information of the terminal device is $C = \{c_1, c_2, \cdots, c_i, \cdots, c_n\}$.

In order to ensure the security of user's identity information and the platform configuration information, the identity authentication expected value and trusted measurement expected value are stored in the COS module of TCDB. In order to repair the platform configuration data of the terminal device, the recover data is stored in the cipher text area of TCDB. In order to verify the credibility of boot process after the operating system running, the log messages which generated by the steps of authentication and trusted measurement are stored in the special area of memory.

• The identity authentication expected value

It is the standard value that judges user's identity information if or not it meets the credibility, represented by u_i. The set is constructed by the identity authentication expected value represented by U.

• The trusted measurement expected value

It is the standard value that judges the platform configuration information of terminal device if or not it meets the credibility, represented by v_i. The set is constructed by the trusted measurement expected value represented by V.

• The recovery data of the platform

Using the recovery data, the TCDB can repair the platform configuration information if the security has problems, represented by rd_i. The set is constructed by the recovery data of the platform represented by RD.

• The log message

The execution steps of the authentication and trusted measurement generate the log messages that are stored in the special area of the memory, represented by msg_i. The set is constructed by the log messages represented by MSG. The application layer can prove the security of the bootstrap by log messages.

When the terminal device power is on, the TCDB verifies the identity information of the user whether it is consistent with the expected value, and then measures whether the platform configuration information does match the expected value and then recovers the data where it does not match. All of these steps generate log messages in order to prove the security of the operating system by the application layer. According to the Definition 1, the boot process meets the trusted boot.

Definition 4: The boot process of the terminal device is trusted
If user's identity and platform configurations do meet the legal credibility of verification, the boot process is considered at terminal device to be credible.

According to the above definitions we describe the execution processes of establishing the trusted root, the trusted measurement and the trusted execution.

1. Establishing the trusted root

During the boot process, the algorithm is executed only once. After users identity information is checked correctly, the BIOS and TCDB are combined together as the trusted root (Fig. 7).

Input:The Set of identities R;
Output:$TRUE$ or $FALSE$;
$FOR\, i = 1\, TO\, |R|\, DO$
$\quad IF\left(sm3(r_i) = u_i\right) THEN$
$\qquad msg_i \rightarrow Memory$
$\quad ELSE$
$\qquad return\ FALSE$
$\quad ENDIF$
$ENDFOR$
$return\ TRUE$

Fig. 7. The algorithm of building trusted root

2. The trusted measurement

The function represents that the recovery data is decrypted by the key. It should be noted that, during the boot process, the algorithm does not execute only once. Before entering the trusted execution stage, the platform configuration information must be restored correctly and measured successfully.

Input:The Set C;
$FOR\, i = 1\, TO\, |C|\, DO$
$\quad h_i = sm3(d_i)$
$\quad IF\left(h_i = v_i\right) THEN$
$\qquad msg_i \rightarrow Memory$
$\quad ELSE$
$\qquad User \rightarrow Trusted\ Three\ Party$
$\qquad Trusted\ Three\ Party \rightarrow User$
$\qquad IF\left(i = 1\right) THEN$
$\qquad\quad sms4\left(key\ \ rd_1\right) \rightarrow c_1$
$\qquad ELSE$
$\qquad\quad rd_i \rightarrow c_i$
$\qquad ENDIF$
$\quad ENDIF$
$ENDFOR$

Fig. 8. The algorithm of building trusted root

3. The trusted execution

As shown in Fig. 6, the communication module decrypts the MBR stored in the TCDB and boot data stored in the disk and move the plain text to the special area of the memory, and then executes them in turn. Ultimately, the OS kernel is loaded and the trusted mechanism is extended to the application layer.

6   Security Analysis and Experiments

6.1   Security Analysis

In security safeguards, TCG regards TPM as a trusted root. At the beginning the CRTM measures the BIOS integrity, and then finishes booting the operating system. But it does not authenticate the user, if the computer is stolen by a malicious person, even if the computer is credible, the data within the machine will be stolen. The scheme combines the BIOS with the TCDB together as the trusted root. In the stage of establishing the trusted root, there is need for "users identity, TCDB and terminal device" certification. In order to ensure that only the inserting, the binding TCDB and providing the correct users identity information, the terminal device can be bootstrap. Even if a malicious person can steal the computer, the data within the machine will not be stolen.

In terms of the integrity being compromised, TCG does not make a specific explanation, and it will affect the user's daily operation. The scheme provides a function that the data of platform can be recovered. In addition, TCDB the recovery data and the boot data are stored in cipher text, and the content of them that is stored in different TCDB vary. It can prevent the attacker from replication.

In the aspect of cryptographic algorithm, TCG releases TPM1.1 and TPM1.2 the use of asymmetric cryptographic algorithm. Even if TPM2.0 specification supports both the asymmetric cryptographic algorithm and symmetric ciphers combination, the ownership and the details of technical implementation are not owned by us. The scheme supports that the cryptographic algorithms in the smart card presented by Chinese scholars, the copyright and details of technical implementation are self-developed.

6.2   Experiments

In order to ignore interactive time between the user's operation and the trusted third party, experiments use the WinDBG and serial lines to build the environment. It can measure the time of the terminal device from powering up to loading the OS kernel.

Table 1. Experimental stages

Stage	Description
First	Do not insert the TCDB, and boot the terminal device from the local disk.
Second	Insert the TCDB, each part of the disk is good.
Third	Insert the TCDB, the MBR is broken.
Forth	Insert the TCDB, the MBR and PBR are broken.
Fifth	Insert the TCDB, and the MBR, PBR and OS Loader are broken.

In order to make the results of experiments more accurate, each of the experiments listed in Table 1 are finished ten times respectively in the target terminal device. The configuration of the target terminal device is Intel i5 processor and 4G of memory. The target operating system is Windows 7 Ultimate, and results of ten times are shown in Fig. 9.

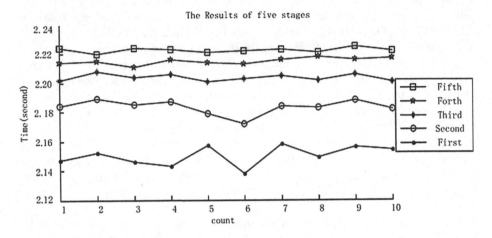

Fig. 9. The test results of five stages

From the extreme values of the Fig. 9, each curve represents one stage. The horizontal ordinate indicates the number of experiments, and the vertical ordinate indicates the execution time. In the first stage, the minimum is about 2.14 s at the sixth time. In the fifth stage, the maximum is about 2.23 s at the ninth time. The difference between maximum and minimum is about 0.09 s, and the user is considered unaffected.

In order to illustrate the usefulness of TCDB deeply, Table 2 lists the average value of each curve in Fig. 8. And by analyzing the specific data, we prove the TCDB practicality.

Table 2. Execution time of experimental stages

Stage	Processing time (ms)
First	2,147
Second	2,184
Third	2,202
Forth	2,214
Fifth	2,219

From the experimental results, the condition that does not recover data of disk spends 37 ms more than the normal bootstrap speed. Otherwise, it will spend 72 ms more than the normal bootstrap speed.

The following conclusions are deduced: in the normal condition that does not recover data of the disk; the TCDB brings 0.3 % overhead to the system. In the worst condition that all the platform configuration information should be recovered, the TCDB brings 0.7 % overhead to the system. Therefore, the TCDB is high availability.

7 Conclusion

In order to improve the security of the operating system at boot time on the basis of hardware and firmware architecture of the smart card and the terminal device, This paper presented a scheme of trusted boot based on the general smart card. In order to ensure the security of the terminal device, the scheme is from powering on to loading the operating system kernel, and then extending the trusted mechanism to the application layer. The scheme has the following features:

1. Practicability

The transformation which bases on the universal smart card with CD-ROM file system and the terminal device does not need the additional hardware. Only some software modules are added.

2. Security

Users identity, TCDB and terminal device are bound together to ensure that the terminal device cannot be started alone.

3. Robustness

In the cases of the boot data of disk is tampered with and the TCDB is lost or damaged, we propose two solutions to ensure the accessibility and security.

4. Trustworthiness

The processes of bootstrap generate log messages that are stored in the special area of memory. When the operating system is started, the application can read the log messages to verify the credibility of the bootstrap.

5. Integrity

The TCDB using the CD file system ensures the data stored in the TCDB cannot be tampered with. And with platform data recovery mechanism, the TCDB can recover the critical data of terminal device automatically.

The scheme is especially suitable for the scenario in which the user brings the laptop computer at work. If the laptop computer is lost, attackers and other users cannot reuse. Ultimately, the scheme improves security. When the user is working, a variety of software are installed and run. Therefore, further work is on the basis of TCDB is to design the trustworthiness platform of the terminal device, and ensures the security of terminal device in running on time.

References

1. Shen, C., et al.: Research and development of trusted computing. J. China Sci. **2**, 139–166 (2010)
2. Shen, C.: Thinking and revelation of cyber space security strategy. J. Fin. Computerizing **6**, 11–13 (2014)
3. TCG PC Client Specific Implementation Specification for Conventional BIOS Specification Version 1.21 Errata Revision 1.00 February 24th, 2012 For TPM Family 1.2; Level 2, 4 (2014). http://www.trustedcomputinggroup.org/
4. UEFI. UEFI specification version 2.3.1 (2011). http://www.uefi.org/specs
5. Huanguo, Z., Zhao, B., et al.: Trusted Computing. Wuhan University Press, Wuhan (2011)
6. Zhibin, H.: The theoretical construction and realization path of state cyberspace security strategy in China. J. China Soft. Sci. **5**, 22–27 (2012)
7. The White House (Washington, USA). International Strategy for Cyberspace-Prosperity, Security, and Openness in a Networked World, 5 (2011). http://www.whitehouse.gov
8. Masti, R.J., Marforio, C., Capkun, S.: An architecture for concurrent execution of secure environments in clouds. In: Proceedings of the 2013 ACM Workshop on Cloud Computing Security Workshop, pp. 11–22. ACM (2013)
9. Danev, B., Masti, R.J., Karame, G.O., et al.: Enabling secure VM-vTPM migration in private clouds. In: Proceedings of the 27th Annual Computer Security Applications Conference, pp. 187–196. ACM (2011)
10. Li, H.-J., Tian, X.-X.: Research of Trust Chain of Operating System. Springer, Heidelberg (2009)
11. Cong, W.-N., Cao, K.: Enabling secure and efficient ranked keyword search over outsourced cloud data. IEEE Trans. Distrib. Syst. **23**, 1467–1479 (2012)
12. Trusted Computing Group. [EB/OL], 4 (2014). http://www.trustedcomputinggroup.org/
13. Tian, J., et al.: Trusted Computing and Trust Management. Science Press, Beijing (2014)
14. Wang, C., Ren, K., Lou, W., et al.: Toward publicly auditable secure cloud data storage services. IEEE Netw. **24**(4), 19–24 (2010)
15. Seo, M.-S., Park, D.-W.: Security measures of personal information of smart home PC. Int. J. Smart Home **7**(6), 227–236 (2013)
16. Wu, L., Li, H., Ren, Y., et al.: Smart card power analysis platform design and implementation. J. Tsinghua Univ. (Sci. Technol.) **10**, 1409–1414 (2012)
17. Cui, J.: Research of Multi-factors Identity Authentication Protocol and Implementation Based on Smart Card. J. East China Normal University 9 (2013)
18. Wang, M.: Research and Design for the Low Power Dual Interface CPU Smart Card Chip. J. Fudan University (2011)
19. Huang, Y.: Researches on Authentication Protocols and Applications in Network Security Systems. Wuhan University, 4 (2010)
20. Zou, D., et al.: Principle and Application of Trusted Computing Technology. Science Press, Beijing (2011)

Efficient Implementation of AND, OR and NOT Operators for ABCs

Antonio de la Piedra[✉]

ICIS DS, Radboud University Nijmegen, Nijmegen, The Netherlands
a.delapiedra@cs.ru.nl

Abstract. In the last few years several practitioners have proposed different strategies for implementing Attribute-based credentials (ABCs) on smart cards. ABCs allow citizens to prove certain properties about themselves without necessarily revealing their full identity. The Idemix ABC is the most versatile ABC system proposed in the literature, supporting peudonyms, equality proofs of representation, verifiable encryption of attributes and proving properties of attributes via AND, NOT and OR operators. Recently, Vullers et al. and De La Piedra et al. addressed the implementation of the selective disclosure operations, pseudonyms and multi-credential proofs such as equality proofs of representation. In this manuscript, we present implementation strategies for proving properties of user attributes via these operators and show how to combine them via external and internal commitment reordering.

Keywords: Attribute-based credentials · Smart cards

1 Introduction

Our everyday life is full of situations where we must identify ourselves. This is exemplified where we buy alcohol, cigarettes or other type of adult goods. In such process, we usually rely on our IDs in order to show that our age is consistent with the current legislation. However, in most of those operations we do not need to reveal our full identity. ABCs solve these privacy breaches by enabling users to reveal or hide the set of attributes that represent their identity according to the real need of the identification process. In so doing, the usual identification operation is replaced by an authorization according to the restricted set of attributes that are asked for. ABCs generally consist of a set of signed attributes that through certain cryptographic primitives can be used for authentication while tracing is avoided as well as ensuring that nobody can reuse the credential attributes or have access to them. Modern anonymous credential systems such as Idemix [8] and U-Prove [4] rely on blind and randomizable signatures in combination with proofs of knowledge [13]. While some practitioners have proposed several implementations of ABCs [3,19], the IRMA card[1] is the

---

[1] https://www.irmacard.org.

© Springer International Publishing Switzerland 2016
M. Yung et al. (Eds.): INTRUST 2015, LNCS 9565, pp. 183–199, 2016.
DOI: 10.1007/978-3-319-31550-8_12

only open-source and practical implementation of Idemix. In this paper we rely on the current version of the card.

The main operation of ABCs is the selective disclosure. A user can reveal a reduced set of her attributes according to a presentation policy sent by a verifier [6]. However, in certain cases it can be useful to prove relations across attributes of the same credential. For instance, a restricted service can enforce an access control based on the ownership of an attribute a OR another one b. Moreover, it can ask if the cardholder is not owning a special type of attribute e.g. one that describes that her age is NOT higher than 18. All these operations are related to the AND, OR and NOT operators introduced by Camenisch et al. in [7]. In this manuscript, we address their implementation on constrained devices.

In the next section, we describe the work of other practitioners who implemented ABCs on smart cards and relate their performance figures with our work. In Sect. 3 we describe the main building blocks of ABCs. In Sect. 4, we sketch out the internals of the IRMA card. In Sects. 5 and 6 we present our strategies for executing complex proofs based on the AND, OR and NOT operators. We describe our results for combining them in Sect. 7. Finally, we end in Sect. 8 with some conclusions.

2 Related Work

Bichsel et al. [3] presented in 2009 the first implementation of Idemix based on Java Card (7.4 s, 1,280 bit RSA modulus) solely based on the selective disclosure of one attribute. These results preceded the design of Sterckx et al. [19] (4.2 s, 1,024 bit RSA modulus). Using the MULTOS platform, Vullers et al. presented an implementation of the issuing and verification operations of Idemix using credentials of 5 attributes (1–1.5 s, 1,024 bit RSA modulus). De La Piedra et al. [18] proposed the implementation of larger credentials, pseudonyms (1,486.51 ms) and multi-credential proofs (2,261.19 ms) on the same platform using a PRNG and variable reconstruction in RAM relying on the implementation of Vullers et al [18].

Contribution. In this manuscript, we present strategies for executing OR, NOT and AND operators over credentials based on prime-encoded attributes as described by [7]. Relying on the AND operator we found the limit of the amount of attributes we can issued in the target device[2] is 44. We always can perform this operation using attributes of different lengths in less than 2.7 s whereas issuing more than 5 attributes using traditional attributes requires more than 3 s [20]. This suggests that issuing prime-encoded attributes can be accompanied by the computation of pseudonyms. On the other hand, we observed that the

---

[2] Our performance figures have been extracted relying on a MULTOS ML3-R3-80K smart card using the SCM Microsystems SCL011 reader in a Intel Core i5-3230M CPU clocked at 2.60 GHz running Debian Linux 3.13.6-1, python 2.7.6, python-pyscard 1.6.12.1-4 and CHARM 0.43 [2].

verification of attributes using this operator is only optimal when none of the attributes are revealed (1.2 s, Sect. 7.1.3). We also present the implementation of the NOT operator using three approaches for solving the required Diophantine equation. Using credentials of the same size of the IRMA card we can perform the NOT operator in 1,974.96 ms with precomputatoin and between 2,016.41 and 2,135.53 ms using the extended Euclidean algorithm. We can perform the OR operator using the same type of credentials in 1,885.96 ms. Finally, we propose the internal and external reorganization of commitments for making it possible the combination of this operators: AND $\wedge$ NOT (2,201.90 ms), AND $\wedge$ OR (1,924.5 ms), NOT $\wedge$ OR (2,122.20 ms) and AND $\wedge$ NOT $\wedge$ OR (2,252.60 ms). By performing the proposed reorganizations of commitments we obtained reductions between 170.10 ms and 644.60 ms and between 9 · 74 and 1 · 74 bytes savings in RAM. Our results suggest that is actually possible to execute complex proofs of knowledge on embedded devices in reasonable times for on-line settings. Moreover, our performance figures are consistent with the current results in the literature [18,20].

3   Preliminaries

The IRMA card relies on a subset of the Idemix specification [20]. In this section, we describe the fundamentals of private ABCs and their main cryptographic blocks: non-interactive commitment schemes, blind signatures and zero-knowledge proofs. In this respect, we present the Camenisch-Lysyanskaya (CL) digital signature [10], the Fujisaki-Okamoto commitment scheme [16] and the Idemix ABC [8].

Non-interactive Commitment Schemes. These constructions are utilized in Idemix for committing to secret values during the issuing and verification operations. In so doing, one of the parties proves the knowledge of a committed value such as an attribute that is not revealed. Typically, a commitment scheme consists of two stages: commit and reveal i.e. a value x that is received as an input in the first stage will be revealed during the second one. Idemix relies on the Fujisaki-Okamoto commitment [16] scheme, which is statistically hiding and computationally binding when factoring is a hard problem. Given an RSA special modulo n, $h \in QR_n$ and $g \in< h >$, the commitment function for an input x, and random value $r \in Z_n$ is computed as $g^x h^r \bmod n$.

Zero-Knowledge Proofs. In a proof of knowledge, a verifier is convinced that a witness w satisfies a polynomial time relation R only known by the prover. If this is performed in a way that the verifier does not learn w, this is called a zero-knowledge proof of knowledge. Damgård proved that is possible to generate zero-knowledge protocols via sigma protocols [14]. In Idemix, the typical three movement of sigma protocols (commitment, challenge and response[3]) is transformed into a Non Interactive Proof of Knowledge (NIZK) via the Fiat-Shamir heuristic [15] in the random oracle model. A variety of zero-knowledge protocols are utilized in Idemix. For instance, proofs of knowledge of discrete logarithm representation modulo a composite are used during issuing and verification [16].

The CL Digital Signature Scheme. The CL signature scheme is the main block of Idemix [10]. It provides multi-show unlinkability via the randomization of the issued signature. This signature is secure under the *Strong RSA* assumption. A CL signature is generated (Gen) by a certain issuer according to her public key $(S, Z, R_0, R_1, ..., R_5 \in QR_n, n)$ using its secret key (p, q). For instance, a CL signature over a set of attributes $(m_0, ..., m_5)$ is computed by selecting A, e and v s.t. $A^e = ZR_0^{-m_0} R_1^{-m_1} R_2^{-m_2} R_3^{-m_3} R_4^{-m_4} R_5^{-m_5} S^{-v}$ mod n. Then, a third party can check the validity of the signature by using the issuer's public key and the triple (A, e, v) as $Z \equiv A^e R_0^{m_0} R_1^{m_1} R_2^{m_2} R_3^{m_3} R_4^{m_4} R_5^{m_5} S^v$ mod n (Verify).

Private ABC Systems. In private ABCs systems [12], the users remain anonymous and are only known by their pseudonyms. They consist of organizations that issue and verify credentials so a user can prove its identity to a verifier while the issuer remains oblivious. In Idemix, this is performed via the multi-show unlinkability property of the CL digital signature scheme. Avoiding the transference of credentials between users is enforced using a secret key that is only known to the user and not by the system (namely, a master secret m_0 in Idemix). In this system, there are two main protocols: issuing (or GrantCred [9]) and verification (or VerifyCred [9]). In the first one, a certain cardholder performs a protocol for signing a committed value, for instance, a set of attributes that represent her identity e.g. $m_0, ..., m_l$ for l attributes. At the end of the protocol, she receives a signature σ whereas the signer did not learn anything about $m_0, ..., m_l$. On the other hand, the verification operation serves for proving the knowledge of a signature over a committed value, for instance a set of attributes and the master secret m_0 of the user for a pair cardholder/verifier. This protocol enables the possibility of using policies (see for instance [5]), i.e. a list of attributes or conditions in a certain credential that must be fulfilled during an authentication operation[4].

---

[3] In the first stage, the prover sends to the verifier a commitment message t or t_value. In the second move, the verifier sends to the prover a random challenge message c. Finally, the last message sent by the prover includes a response value or s_value.

[4] For instance, an empty proof of possession over a set of attributes $(m_0, ..., m_5)$ is represented using the Camenisch-Staedler notation [11] as: NIZK: $\{(\varepsilon', \nu', \alpha_0, ..., \alpha_5) : Z \equiv \pm R_0^{\alpha_0} R_1^{\alpha_1} R_2^{\alpha_2} R_3^{\alpha_3} R_4^{\alpha_4} R_5^{\alpha_5} A^{\varepsilon'} S^{\nu'} \mod n\}$ being the Greek letters (ε', ν') and $(\alpha_0, ..., \alpha_5)$ the values of the signature and the set of attributes proved in zero knowledge and not revealed.

4 The IRMA Card

IRMA supports up to 5 attributes by credential and relies on 1,204 special RSA modulus for performance reasons[5]. IRMA is based on the MULTOS card. Particularly, our target device is the ML3-80K-R1 version. It is based on the SLE 78CX1280P chip by Infineon[6]. This processor, clocked up to 33 MHz, provides an instruction set compatible with the Intel 8051 and hardware accelerators for ECC, RSA, 3DES and AES.

Issuing in IRMA. Issuing in Idemix is related to the generation of a CL blind signature over the attributes of the cardholder. In so doing, the issuer cannot extract the master secret m_0 of the cardholder and the generated tuple (A, e, v) remains hidden too. However, in IRMA the cardholder's attributes are never revealed to the issuer.

Table 1. Message flow for issuing a CL signature over a set of attributes (I: Issuer, C: Cardholder)

Common inputs: The public key of the issuer $(S, Z, R_0, R_1, ..., R_5 \in QR_n, n)$, the number of attributes that are issued.
Cardholder inputs: The credential involved and its set of attributes $m_0, ..., m_5$.
Issuer inputs: The private key of the issuer p, q.

Protocol: The cardholder first proves the knowledge of the representation of U as NIZK: $\{(\nu', \alpha_0) : U \equiv \pm S^{\nu'} R^{\alpha_0} \mod n\}$. Then, the issuer proves the knowledge of $1/e$.

1. $I \to C$: The user chooses a random nonce $n_1 \in_R \{0,1\}^{l_\emptyset}$ and sends it to the cardholder.
2. $C \to I$: The cardholder computes the commitment $U = S^{\nu'} R^{m_0} \mod n$ with $v' \in \pm\{0,1\}^{l_n+l_o}$. It proves in zero knowledge m_0 with a randomizer α_0, v'. Then ,it generates the s_values for $\hat{v}'$, $\hat{r}$ together with the challenge c with the context, U and the nonce n_1. Finally, it sends to the issuer $n_2 \in_R \{0,1\}^{l_o}$.
3. $I \to C$: It verifies the NIZK as $\hat{U} = U^{-c}(S^{\hat{v}'} R_0^{\hat{\alpha}_0}) \mod n$. Then, it proves the knowledge of $\frac{1}{e}$ via random values $e, \tilde{v}, v''$. It computes the commitments $Q = Z\,U^{-1}S^{-v''} \mod n$, $A = Q^{e^{-1} \mod p'q'} \mod n$. It sends (A, e, v'') and creates the proof NIZK: $\{(1/e) : A \equiv \pm Q^{e^{-1}} \mod n\}$. It computes $\tilde{A} = Q^r \mod n$ for $r \in_R Z_{p'q'}$, and obtains the challenge c' as the hash of the context, Q, A, n_2 and $\tilde{A}$. It computes $S_e = r - c' \cdot e^{-1} \mod p'q'$ and sends A, e, v'', S_e, c' to the cardholder.
4. C: Computes $v = v'' + v'$ and verifies (A, e, v) as $Q = Z\,S^{-v}R_0^{m_0} \mod n$ with $\hat{Q} = A^e \mod n$.

---

[5] As described in [18], the attributes are represented as $l_m = 256$ bits. The rest of parameters are set as $l'_e = 120$ (size of the interval where the e values are selected), $l_\emptyset = 80$ (security parameter of the statistical ZKP), $l_H = 256$ (domain of the hash function in the Fiat-Shamir heuristic), $l_e = 504$ (size of e), $l_n = 1,024$ (size of the RSA modulus) and $l_v = 1,604$ bits (size of v).

[6] http://www.infineon.com/dgdl/SPO_SLE+78CX1280P_2012-07.pdf? folderId=db3a304325afd6e00126508d47f72f66&fileId= db3a30433fcce646013fe3d672214ab8 (Accessed 27 February 2015).

Issuing requires two NIZK (Table 1). In IRMA, the issuing part of Idemix mimics the Cardholder-Issuer interaction as a set of states: ISSUE_CREDENTIAL, ISSUE_PUBLIC_KEY, ISSUE_ATTRIBUTES, ISSUE_COMMITMENT, ISSUE_COMMITMENT_PROOF, ISSUE_CHALLENGE, ISSUE_SIGNATURE and ISSUE_VERIFY. The first, state ISSUE_CREDENTIAL, puts the card in issuance mode, sends the identifier of the credential that will be issued and the context of the operation. Then, during the ISSUE_PUBLIC_KEY state, the card accepts the public key of the issuer: $n, S, Z, R_0, ..., R_5$. The attributes to be issued are sent to the card in the ISSUE_ATTRIBUTES state. The rest of the states are related to the execution of the two NIZK. During ISSUE_COMMITMENT the cardholder receives the nonce n_1, it computes U and returns it. Then, in ISSUE_COMMITMENT_PROOF, the required values for proving the knowledge of m_s in U: $c, \hat{v}', \hat{s}$ are generated. In ISSUE_CHALLENGE, it sends n_2. During the ISSUE_SIGNATURE mode, the issuer constructs the blinded CL signature and sends to the card the partial signature (A, e, v''). Finally, in ISSUE_VERIFY the card verifies the signature using the values sent the verifier (c, S_e).

We can model the latency of the issuing process in the IRMA card by representing the time required for performing the operation described in Table 1 as $T_{issuing}(n)$ where n is the number of attributes that will be issued in a certain credential. This latency would be result of summing up the time required for getting the public key of the issuer, adding the computation of the involved proofs and the process of obtaining and verifying the signature:

$$T_{issuing}(n) = T_{sel\_cred} + \sum_{i=n,S,Z,R_i} T_{get\_PK}(i) +$$
$$\sum_{i=1}^{n} T_{get\_attr}(i) + T_{gen\_commitment} + \sum_{i=c,\hat{v}',\hat{s}} T_{gen\_proof}(i) + \qquad (1)$$
$$\sum_{i=A,e,v''} T_{get\_signature}(i) + T_{verify}(n)$$

From this model, we know that there are only two operations that depends on the number of attributes issued that are part of a certain credential: $T_{get\_attr}(i)$ and $T_{verify}(n)$. That would mean that in order to optimize the overall latency of $T_{issuing}(n)$ there are two strategies: (1) reduce the number of attributes that are part of the credential (we analyze this aspect in Sect. 5.1) and (2) reduce then number of operations in the verification part of the proof, which is already implemented on the IRMA card where the second proof is optionally verified for reducing the computational complexity of the operation.

Verification in IRMA. When the card receives a verification request, it changes its initial state to PROVE_CREDENTIAL. Then, it acquires a presentation policy with the description of the attributes that must be revealed. Then, the card performs the operations depicted in Table 2 (PROVE_COMMITMENT). Afterwards, the card changes its working state to PROVE_SIGNATURE. In this state, the verifier can request the randomized tuple $(A', \hat{e}, \hat{v}')$. Finally, the card switches to PROVE_ATTRIBUTE, where the verifier is allowed to request the set of revealed and hidden attributes related to the proof. This set of states is mapped to the three moves described in Table 2.

Table 2. Message flow for proving the ownership of a CL signature over a set of attributes (V: Verifier, C: Cardholder)

Common inputs: The public key of the issuer $(S, Z, R_0, R_1, ..., R_5 \in QR_n, n)$, the presentation policy i.e. those attributes that must be revealed $m_i \in A_r$ w.r.t. to those that can be hidden $m_i \in A_{\bar{r}}$.
Cardholder inputs: The credential involved and its set of attributes $m_0, ..., m_5$, randomizers and the tuple (A, e, v) over $m_0, ..., m_5$.
Verifier inputs: $A', v', m_i \in A_r, m_i \in A_{\bar{r}}$

Protocol: The (cardholder, verifier) pair perform the following NIZK where the cardholder proves the ownership of a CL signature over $m_0, ..., m_5$ and reveals, for instance m_1:
NIZK: $\{(\varepsilon', \nu', \alpha_0, \alpha_2, \alpha_3, \alpha_4, \alpha_5) : ZR_1^{-m_1} \equiv \pm R_0^{\alpha_0} R_2^{\alpha_2} R_3^{\alpha_3} R_4^{\alpha_4} R_5^{\alpha_5} A^{\varepsilon'} S^{\nu'} \mod n\}$.

1. $V \rightarrow C$: The verifier sends a fresh nonce n_1 to the verifier together with a presentation policy that must be fulfilled.
2. $C \rightarrow V$: Using the tuple (A, e, v) over $m_0, ..., m_5$, it randomizes the signature by generating $r_A \in_R \{0, 1\}^{l_n + l_\emptyset}$ and obtaining the new tuple (A', e', v') as $A' = AS^{r_A}$ $\mod n$, $v' = v - er_A$ and $e' = e - 2^{l_e - 1}$. Afterwards, it generates the randomizers $\tilde{e} \in_R \pm \{0, 1\}^{l'_e + l_\emptyset + l_H}$, $\tilde{v}' \in_R \pm \{0, 1\}^{l_v + l_\emptyset + l_H}$, and $\tilde{m}_i \in_R \pm \{0, 1\}^{l_m + l_\emptyset + l_H} (i \in A_{\bar{r}})$ and computes the commitment (t-value) $\tilde{Z} = A'^{\tilde{e}} (\prod_{i \in A_{\bar{r}}} R_i^{\tilde{m}_i}) S^{\tilde{v}'}$. Then, it generates the challenge c by hashing the context, the t-value and the nonce. It generates the response values (s-values) $\hat{e} = \tilde{e} + ce'$, $\hat{v}' = \tilde{v}' + cv'$ and $\hat{m}_i = \tilde{m}_i + cm_i (i \in A_{\bar{r}})$. Finally, It sends the challenge c, the common value A', the response values and $m_i \in A_r$ to the verifier.
3. V: Verifies if the proof is correct via the issuer public key by computing $\hat{Z} = (Z A'^{-2^l e - 1} \prod_{i \in A_r} R_i^{-m_i})^{-c} (A')^{\hat{e}} (\prod_{i \in A_{\bar{r}}} R_i^{\hat{m}_i}) S^{\hat{v}'})$ and checking if c equals the hash of A', $\hat{Z}$ and n_1.

As described in [18] the latency of the verification operation can be modeled first according to the number of attributes per credential together with the number of attributes that are revealed (r) or hidden.

$$T_{verify}(n, r) = T_{sel\_cred} + T_{gen\_commit}(n, r)$$
$$+ \sum_{i=A,e,v} T_{get\_sig}(i) + \sum_{i=1}^{n} T_{get\_attr}(i) \tag{2}$$

The time the **PROVE_CREDENTIAL** state requires is represented by T_{sel_cred}. Further, $T_{gen\_commit}(n, r)$ represents **PROVE_COMMITMENT**. Finally, $T_{get\_sig}(i)$ is related to the **PROVE_SIGNATURE** state whereas $T_{get\_attr}(i)$ represents the **PROVE_ATTRIBUTE** state.

We rely on the PRNG proposed by De La Piedra et al. in [18] for recomputing the associated pseudorandomness of the proofs. That approached made it possible to increase the number of attributes per credential by recomputing the $\tilde{m}_i$ values. In so doing, it is possible to generate the associated pseudorandomness during the generation of the t-values and obtain, on the fly, the same sequence while generating the s-values by resetting the PRNG as described in [18] e.g. $init_{PRNG}() \Rightarrow \tilde{m}_i \Rightarrow reset_{PRNG}() \Rightarrow \tilde{m}_i$.

5   Performance Evaluation of AND, OR and NOT Operators

The main operation of Idemix is the modular exponentiation. This operation is related to the number of attributes that a certain cardholder hides in an operation. In [7], Camenisch et al. proposed encoding the user attributes as prime numbers, reducing the overall number of modular exponentiations to 2. In so doing, they only utilize a base R_1 for encoding all the attributes as product $m_t = \prod_{i=1}^{l} m_i$ for l attributes. This encoding technique enables the possibility of performing selective disclosure (namely, using the AND operator), proving the absence of an attribute in a certain credential (NOT operator) and the possibility that one or more attributes are presents in m_t s.t. $R_1^{m_t}$ via the OR operator. This encoding technique is useful where the number of possible values in an attribute is restricted to only some e.g. masculine or feminine, and each possibility has associated a prime number.

We rely on the PRNG described in [18] for making it possible the execution of these proofs. Moreover, we introduce two techniques (internal and external commitment reorganizations) for reducing the amount of required exponentiations and the RAM required for storing the respective commitment in each step. External commitment reorganizations make it possible enabling the chaining of several proofs using the AND, NOT and OR operators in tandem. The internal reorganization of commitments means reordering the computations of commitments of a certain proof in order to save the computation time and the amount of utilized RAM.

5.1   The AND Operator

This operator performs the selective disclosure of these attributes by proving that a certain value m_i (which can be one attribute or a product of several ones) divides m_t. In this respect, proving that a certain attribute m_1 belongs to m_t is represented in zero knowledge as NIZK: $\{(\varepsilon', \nu', \alpha_0, \alpha_1) : Z \equiv \pm R_0^{\alpha_0} (R_1^{m_1})^{\alpha_1} A^{\varepsilon'} S^{\nu'} \mod n\}$. In addition, the commitments $C = Z^{m_t} S^r \mod n$, $\tilde{C} = (Z^{m_1})^{\tilde{m}_h} S^r \mod n$ and $\tilde{C}_0 = Z^{\tilde{m}_t} S^{\tilde{r}} \mod n$ must be computed, where $m_h = m_t / m_i$ and m_i consists of the product of attributes that are revealed (in this case $m_i = m_1$).

In this case, the PRNG would compute the following sequence: $init_{PRNG}() \Rightarrow \tilde{m}_i \Rightarrow \tilde{m}_h \Rightarrow \tilde{r} \Rightarrow r \Rightarrow \tilde{m}_t \Rightarrow reset_{PRNG}() \Rightarrow \tilde{m}_i \Rightarrow \tilde{m}_h \Rightarrow \tilde{r}$. Otherwise, not revealing any attribute, that is, only proving the ownership of the signature would be represented as NIZK: $\{(\varepsilon', \nu', \alpha_0, \alpha_1) : Z \equiv \pm R_0^{\alpha_0} R_1^{\alpha_1} A^{\varepsilon'} S^{\nu'} \mod n\}$. This requires two exponentiations with independence of the number of attributes hidden.

We can apply the internal organization of commitments. For instance, in the computation of the AND proofs we need to commit to the m_t value, i.e. the first attribute of the first base as $C = Z^{m_t} S^r \mod n$. However, the next commitment requires the computation of the S^r again as $\tilde{C} = (Z^{m_1})^{\tilde{m}_h} S^r$. In order to avoid recomputing S^r, we can proceed by reordering all the computations and reuse this value from the last commitment. In this case, the order of computations would be (1) $Z^{m_t}S^r$, (2) $[S^r](Z^{m_1})^{\tilde{m}_h}$ by leaving the result S^r in RAM and proceeding with the next multiplication. This resulted in an speed up of 78 ms per operation.

Issuing Prime-Encoded Attributes. Since the number of bases (and modular exponentiations) is reduced to the number of attributes in the credential to 2, we can compare the performance of the Idemix issuing operation using both prime-encoded and traditional attributes [20]. In this respect, it is expected that issuing prime-encoded attributes could reduce the latency associated to $T_{issuing}(n)$ as the number of attributes increase (Sect. 4). In the IRMA card, only 5 attributes w.r.t. the bases $R_0, ..., R_5$ are used. On the other hand, we can store any number of prime-encoded attributes s.t. the only limitation would be the prime size. Hence, we can compare how the issuing operation in Idemix scales and observe how many attributes we can issue in the limit case of IRMA (5 attributes) [20].

We rely on the following methodology. First, we set a limit of 50 attributes per credential. Then, the only restriction is that $|m_t|$ cannot be greater than the $l_m = 256$ bit limitation according to the Idemix specification. Hence, we create the following cases: (1) one possibility per attribute: we rely on the first 50 primes, (2) 10 possibilities per attribute: we rely on the first 500 primes, (3) 100 possibilities per attribute, we rely on the first 5,000 primes, (4) 1,000 possibilities per attribute, we rely on the first 50,000 primes. We select attributes from the list of the first 50 primes, 500 primes, 5,000 primes, 50,000 primes and so on in order to construct our credentials w.r.t. the m_t exponent for the base R_1 as $\prod_{i=0}^{l} m_i$ for $l = 50 - 1$.

What we want to know is for each case, what is the maximum number of attributes per credential we can store according to l_m. Then, we create a list of primes according to its possibilities in each case when 50, 500, 5,000, 50,000

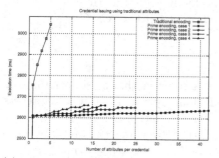 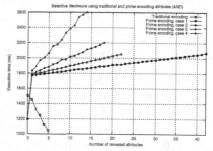

(a) Performance of issuing attributes via prime-encoded and traditional attributes

(b) Performance of selective disclosure using both types of encoding

Fig. 1. Issuing and verifying prime-encoded attributes

primes are involved. We can randomly choose prime numbers from that list and construct our credentials from 1 to 50 attributes, stopping when $|m_t| \geq 256$ bits. After repeating this experiment 100 times for each case, we obtained the approximate maximum number of attributes: 44 attributes (case 1), 25 attributes (case 2), 18 attributes (case 3) and 14 attributes (case 4).

In relation to Fig. 1 (a), we have used the total number of bits for encoding the attributes for each case in order to obtain a fair comparison. As depicted, it is possible to not cross the 3 s margin of issuing traditional attributes and at the same time issue 44, 22, 18 and 14 attributes of different lengths under the 2.7 s limit. This could mean that it would be possible to also associated a pseudonym or several pseudonyms during the issuing of that credential and still maintain a decent performance in comparison to the utilization of traditional attributes but issuing from 3 to 8 times more attributes [20].

Verification of Prime-Encoded Attributes. Due to the computation of the $C, \tilde{C}_o$ and $\tilde{C}$ commitments together with the two extra response values, revealing attributes via the AND operator undermines any speed up in comparison to the issuing operation relying on traditional attributes. We take the limit case of IRMA (5 attributes) and compare it with verifying and hiding prime encoded attributes in Fig. 1 (b). Hiding attributes is computationally more expensive using traditional attributes in comparison to prime-encoded ones whereas hiding attributes only has a constant performance related to prove the ownership of m_0 and m_t w.r.t. $R_0^{m_0} R_1^{m_t}$. Hence, only proving the ownership of a CL signature over a set of attributes without revealing any only requires 1,198.21 ms. In contrast, revealing all the attributes requires the computation of C, $\tilde{C}$ and $\tilde{C}_0$.

Besides, the cost of this operation is related to the computation of the Z^{m_r} exponentiation w.r.t. of m_r as the product of the cardholder's attributes that are revealed together with the product itself ($T_{gen\_commit}(n, r)$). Therefore, it is expected that the AND operator increases the computation time as the number of attributes are revealed at a speed related to the primes utilized (Fig. 1 (b)). In this respect, an alternative for reducing the latency of $T_{gen\_commit}(n, r)$ is to precompute a restricted set of combinations for revealing attributes Z^{m_r} and store them in ROM so $T_{gen\_commit}(n, r)$ is constant w.r.t. $C, C_0, \tilde{C}$.

5.2   The NOT Operator

By using prime-encoded attributes it is possible to prove that an m_i attribute or set of attributes do not belong to m_t. This is done by showing that the integers x, y exists w.r.t. the following linear Diophantine equation $x \cdot m_t + y \cdot m_i = 1$.

We prove the ownership of a CL signature over m_0, m_t and the existence of (x, y) via zero knowledge as NIZK: $\{(\varepsilon', \nu', \alpha_0, \alpha_1, \chi, \upsilon, \rho, \rho') : Z \equiv \pm R_0^{\alpha_0} R_1^{\alpha_1} A^{\varepsilon'} S^{\nu'}$ mod n $\wedge$ C $\equiv \pm Z^{\alpha_1} S^{\rho}$ mod n $\wedge$ Z $\equiv \pm C^{\chi}(Z^{m_i})^{\upsilon} S^{\rho'}\}$ mod n. The card must compute the commitments $C = Z^{m_t} S^r$ mod n, $\tilde{C} = C^{\tilde{x}}(Z^{m_i})^{\tilde{y}} S^{\tilde{r}'}$ mod n and $\tilde{C}_c = Z^{\tilde{m}_t} S^{\tilde{r}}$ mod n where $\tilde{r}, \tilde{r}', \tilde{x}, \tilde{y}$ are randomizers [7].

In this case, the critical operation is the computation of the *(a, b)* pairs and how this operation scales for large primes. We propose two type of implementations: (1) precomputing the pairs (x, y) and (2) solving the Diophantine equation on the card. In the first case, given (x, y) for $x \cdot m_i + y \cdot m_t = 1$, m_t has always the same value and there are several combinations for m_i. Those possibilities can be stored in EEPROM if a small number of attributes is utilized. The number of (x, y) pairs that must be stored is related to the number of attributes per credential as $\sum_{i=1}^{l} c_n^i = \sum_{i=1}^{l} \binom{n}{i}$. Hence, for $l = 2$ attributes per credential, we would have to store three (x, y) pairs. In the case of 4 attributes, we store 15 pairs.

The second design is related to the computation of the Extended Euclidean algorithm on the smart card. We can use the instruction PRIM_DIVIDEN from the MULTOS specification that extracts the Euclidean division of two numbers i.e. q and r $(\mathcal{O}(ln^3 N))$ in order to implement it. Other alternative is to use the binary GCD or Stein's algorithm [17]. This algorithm replaces the multiplications and divisions by bit-wise operations. Finally, the Lehmer's algorithm relies on the following idea $(\mathcal{O}(ln^2 N))$. When a and b have the same size, the integer part w of the quotient a/b has one only digit. The goal is to find w while it is small and continue the involved operations via a matrix. The advantage of this method is that a large division is only done when needed, if w is greater than a certain base M.

In order to test the performance of these three algorithms, we have created four possible cases and have extracted performance figures in our target device. The attribute m_i can vary according to the number of attributes that are proved that are not in m_t. Its length will be greater according to the number of possibilities for each credential. In order to obtain an estimation of the computation time of each method on the MULTOS card we take 4 cases. If we take the first 10,000 primes, the numbers consist of 2 to 104,729. We can encode these values using 1 byte to 3 bytes (e.g. 0×019919 in the case of 104,729). However, m_i and m_t can increase according to all the possibilities an attribute can represent together with the number of involved attributes. We take four cases for an implementation based on 5 attributes (m_t) with different possibilities[7]. We rely on credentials of 5 attributes in this case in order to compare the performance of this operation with the selective disclosure via traditional attributes of the IRMA card [20].

---

[7] Thus, for one possibility per attribute, we prove the non-existence of one attribute in m_i. In this case, $m_i = 3$ and $m_t = 5 \cdot 7 \cdot 11 \cdot 13$ (case 1). We consider 10 possibilities per attribute (50 primes). We prove the non-existence of one attribute in m_i. For $m_i = 3$, $m_t = 179 \cdot 181 \cdot 191 \cdot 193$ (case 2). We consider 1,000 possibilities per attribute (i.e. 5,000 primes) and we prove the non-existence of two attributes in m_t for $m_i = 1,999 \cdot 2,161$ and $m_t = 3,323 \cdot 3,253 \cdot 2,897 \cdot 2,999$ (case 3). Finally, we consider 10,000 possibilities per attribute (50,000 primes) and we proof the non-existence of two primes $m_i = 91,387 \cdot 91,393$ in $m_t = 102,461 \cdot 102,481 \cdot 102,497 \cdot 102,499$ (case 4).

Table 3. Performance of GCD using the proposed algorithms

Case	$|m_i|$ (bytes)	$|m_t|$ (bytes)	Euclid (ms)	Stein (ms)	Lehmer (ms)	Extended Euclidean Algorithm (Euclid, ms)
1	1	2	**17.05**	70.62	18.43	**21.51**
2	1	4	**17.52**	131.78	18.92	**21.76**
3	3	6	**37.49**	254.37	38.86	**65.64**
4	5	9	**68.07**	289.58	95.61	**131.47**

The first aspect we notice from Table 3 is that the Stein's variant obtained the worst computational figures for the cases proposed despite it is been based on bit-wise operations, that are suppose to require less time as claimed by Akhavi et al. [1]. This is, however not true for MULTOS. For the maximum length of case 4 (9 bytes), we have measured the latency of all the operations involved: Euclidean division (11.852 ms), comparison (11.047 ms), Boolean and (10.411 ms), right shift (10.634 ms), increment (10.354 ms) and subtraction (10.647 ms). These latencies make this option ill-suited when replacing the Euclidean division by operations that are suppose to require less cycles. On one hand, the Stein's variant requires more control operations and branches and on the other one, bit-wise operations have a similar latency than the Euclidean division. Due to the proprietary nature of the SLE 78CX1280P chip we cannot claim that the Euclidean division is being performed via the hardware accelerator of the target device. Moreover, since MULTOS is based on MEL byte code that is executed in a virtual machine, we cannot be sure that code optimizations (written in C) can result in any speed up. Finally, we are unaware of any side channel analysis (SCA) countermeasures implemented on the card, but there is a possibility that the designers wanted to homogenize the latency of a group of simple arithmetic operations in order to make them indistinguishable.

In the case of the Lehmer's variant, for single-precision values of 32 bits or less, we obtain similar results as the Euclidean algorithm. We believe that due to that when we overcome that value (multi-precision), there are more calls to the operating system for performing bit-wise operations, multiplications and divisions that increase the latency of the algorithm despite this is not expected, whereas in the traditional Euclidean algorithm we are only performing one Euclidean division by step. Moreover, in our target device is not possible to tune the precision and adjust the assembler code since that is then translated into byte codes, executed by the virtual machine.

We have depicted in Table 4[8] the performance figures of the NOT operator for each case. In the precomputation strategy we only show the first case since increasing the length of the operand does not alter the result significantly. On the other hand, the computation of the pairs (x, y) is performed during the

---

[8] We use the following notation in Tables 4, 5 and 6: PRE means precomputation, EUC 1-3 is related to the cases presented in Table 3, RA means Reveal all the Attributes with the exception of the master secret and HA to hide every attribute in the credential.

Table 4. Performance figures of the NOT operator while precomputing the (x, y) pair and relying on the Euclidean algorithm (ms)

Case	T_{sel_cred}	T_{gen_commit}	$T_{get\_sig}(A, e, v)$	$T_{get\_attr}(\hat{m}_0, \hat{m}_t \mid \hat{r}, \hat{r}', \hat{a}, \hat{b}, C)$	Total
NOT PRE	15.203	1,590.11	48.72	214.80	**1,974.96**
NOT EUC (1)	15.503	1,587.93	46.81	259.95	2,016.41
NOT EUC (2)	15.514	1,587.92	15.677, 47.10	260.53	2,017.23
NOT EUC (3)	15.510	1,587.93	46.87	306.28	2,063.63
NOT EUC (4)	15.511	1,587.91	46.85	376.28	2,135.35
OR	113.622	1,476.47	46.08	234.53	1,885.96

computation of $\hat{a}$, adding its latency to $T_{get\_attr}(i)$ (Sect. 2). Thanks to the low latency operation of the PRIM_DIVIDEN primitive we obtained latencies between 2,015.41 and 2,135.35 ms.

5.3   The OR Operator

The utilization of this operator enables the cardholder to prove that an attribute m_i or more attributes encoded as a product can be found in m_t s.t. $m_t = \prod_{i=0}^{l}$ w.r.t. $R_1^{m_t}$. In so doing, we rely on the following fact: given an attribute $m_i \in m_t$, an integer x exists s.t. $x \cdot m_i = \prod_{i=1}^{l} m_i = m_t$. This is proved in zero knowledge as NIZK: $\{(\varepsilon', \nu', \alpha_0, \alpha_1, \chi, \rho, \rho') : Z \equiv \pm R_0^{\alpha_0} R_1^{\alpha_1} A^{\varepsilon'} S^{\nu'} \bmod n \wedge C \equiv \pm Z^{\alpha_1} S^{\rho} \bmod n \wedge C' \equiv \pm C^{\chi} S^{\rho'}\} \bmod n^9$. The card must compute three commitments $C = Z^{m_t} \cdot S^r \bmod n, \tilde{C} = Z^{\tilde{m}_t \cdot S^{\tilde{r}}} \bmod n, \tilde{T} = C^{\tilde{x}} \cdot S^{\tilde{r}_1}$ w.r.t $x = \frac{m_t}{m_i}$ s.t. $R_1^{m_t}$ and $r_1 = -r_0 \cdot x$ where $r, r_0, \tilde{r}, \tilde{r}_0, \tilde{r}_1, \tilde{m}_t, \tilde{x}$ are randomizers. The first obstacle for implementing this primitive was to over come the lack of support of signed arithmetic on the card. This means creating wrappers over the multiplication and addition operations supporting sign extensions due to the computation of $r_1 = r_0 \cdot x$. Afterwards, the operation is performed in two's complement. By using the RAM reductions achieved thanks to the PRNG described in [18] and executing all the two's complement operations in RAM, we cold reduce the computational time of $r_1 = -\rho_0 \cdot \chi$ from 495.530 ms to 90.260 ms.

We have depicted in Table 4 the performance figures of case 1, described in Sect. 5.1. We can compute this operation withing 1,885.96 ms. Since this operation scales with m_i at the same pace of the AND, OR operators without the m_t product. Since the commitments utilized only involved two multi modular exponentiations, we can obtain a reduction of 1,974.96 - 1,885.96 (89) ms in comparison to the NOT operation.

---

[9] In this manuscript we only address the first version of this NIZK described in [7] and leave the second one beyond the scope of this work due to the computation limitations of our target device.

Table 5. Estimation of the performance obtained by the combination of operators for prime-encoded credentials (5 attributes)

Combination	Cases	Performance (ms)	Performance after optimization (ms)
AND ∧ NOT	RA, PRE	2,485.3	2,201.9
AND ∧ NOT	RA, EUC1	2,506.9	2,223.4
AND ∧ NOT	RA, EUC2	2,507.1	2,223.7
AND ∧ NOT	RA, EUC3	2,551.0	2,267.5
AND ∧ NOT	RA, EUC4	2,616.8	2,333.4
AND ∧ OR	RA, C1	2,247.7	1,924.5
NOT ∧ OR	PRE, C1	2,292.3	2,122.2
NOT ∧ OR	EUC1, C1	2,397.9	2,227.8
NOT ∧ OR	EUC2, C1	2,365.2	2,195.1
NOT ∧ OR	EUC3, C1	2,409.1	2,238.9
NOT ∧ OR	EUC4, C1	2,474.9	2,304.8
AND ∧ NOT ∧ OR	RA, PRE, C1	2,897.1	2,252.6
AND ∧ NOT ∧ OR	RA, EUC1, C1	2,918.6	2,274.1
AND ∧ NOT ∧ OR	RA, EUC2, C1	2,918.9	2,274.3
AND ∧ NOT ∧ OR	RA, EUC3, C1	2,962.8	2,318.2
AND ∧ NOT ∧ OR	RA, EUC4, C1	3,028.6	2,384.0

5.4 Combination of Operators for Prime-Encoded Credentials

It can be useful to prove certain properties of a prime-encoded credential by utilizing a group of these operators. For instance, one could prove that an attribute a is in m_t s.t. $R_1^{m_t}$, b is NOT AND c OR d could be present. In so doing, it can be possible to perform some degree of commitment reorganization (i.e. external reorganization) in order to optimize the computation of the required commitments and response values.

Given the AND, NOT and OR operators, we consider the following combinations in order to obtain the best combination and estimate its performance. First, we discuss AND ∧ NOT. In the AND proof we always to commit to m_t as $C = Z^{m_t} \cdot S^r$ in order to prove that a certain m_1 can divide m_t afterward and utilize the $\tilde{m}_t, \tilde{r}$ randomizers for proving the ownership of m_t as $\tilde{C}_0 = Z^{\tilde{m}_t} \cdot S^{\tilde{r}}$. The response values $\hat{m}_t, \hat{r}$ are created. The NOT operator follows a similar approach for proving the ownership of m_t in the case of the C and $\tilde{C}_c$ commitments (Sect. 5.2). Hence, when proving both presence an absence of attributes one can avoid computing these two commitments and their response values twice. Moreover, in the case of AND we can apply internal commitment reorganization. Then, in AND ∧ OR, the OR operator (Sect. 5.3) proves the ownership of m_t as $C = Z^{m_t} S^r$ and generates $\tilde{C}$ as the AND and NOT operator as well as the response values for $\hat{m}_t, \hat{r}$. This means that it can be computed only one time

Table 6. RAM savings by recomputing the pseudorandomnes in each primitive

Operator	PRNG sequence	No. of commitments	Required RAM (bytes)	RAM saved (bytes)
AND RA	$\hat{m}_0, \hat{m}_h, \hat{r}, r, \hat{m}_t \| [\_], \hat{m}_0, \hat{m}_h, \hat{r}, [\_], [\_], \hat{m}_t$	4	222	148
AND HA	$\hat{m}_t, \hat{m}_0 \| \hat{m}_t, \hat{m}_0$	1	74	-
AND $\wedge$ Nym RA	$\hat{m}_0, \hat{m}_h, \hat{r}, \hat{r}', r, \hat{m}_t \| [\_], \hat{m}_0, \hat{m}_h, \hat{r}', \hat{r}, [\_], [\_], \hat{m}_t$	5	370	296
AND $\wedge$ Nym HA	$\hat{m}_t, \hat{m}_0, \hat{r}' \| \hat{m}_t, \hat{m}_0, \hat{r}'$	2	148	74
NOT	$\hat{b}, \hat{a}, \hat{r}', \hat{m}_t, \hat{r}, \hat{m}_0, r \| \hat{b}, \hat{a}, \hat{r}', \hat{m}_t, \hat{r}, \hat{m}_0$	4	444	370
OR	$r, \hat{m}_t, \hat{r}, \hat{r}_1, \hat{x}, \hat{m}_0 \| [\_], \hat{m}_t, \hat{r}, \hat{r}_1, \hat{x}, \hat{m}_0$	3	370	296
AND RA $\wedge$ NOT	$\hat{b}, \hat{a}, \hat{r}', \hat{m}_t, \hat{r}, \hat{m}_0, r, \hat{m}_h \| \hat{b}, \hat{a}, [\_], \hat{m}_t, \hat{m}_0, [\_], \hat{m}_h$	6	370	296
AND HA $\wedge$ NOT	$\hat{b}, \hat{a}, \hat{r}', \hat{m}_t, \hat{r}, \hat{m}_0 \| \hat{b}, \hat{a}, [\_], \hat{m}_t [\_], \hat{m}_0$	5	296	222
AND RA $\wedge$ OR	$\hat{m}_0, \hat{m}_h, \hat{r}, r, \hat{m}_t, \hat{r}_1, \hat{x} \| \hat{m}_0, \hat{m}_h, \hat{r}, [\_], \hat{m}_t, \hat{r}_1, \hat{x}$	5	444	370
AND HA $\wedge$ OR	$r, \hat{m}_t, \hat{r}, \hat{r}_1, \hat{x}, \hat{m}_0 \| [\_], \hat{m}_t, \hat{r}, \hat{r}_1, \hat{x}, \hat{m}_0$	3	370	296
NOT $\wedge$ OR	$\hat{b}, \hat{a}, \hat{r}', \hat{m}_t, \hat{r}, \hat{m}_0, \hat{r}', \hat{r}, \hat{r}_1, \hat{x} \| \hat{b}, \hat{a}, \hat{r}', \hat{m}_t, \hat{r}, \hat{m}_0, \hat{r}', [\_], \hat{r}_1, \hat{x}$	5	666	592
AND RA $\wedge$ NOT $\wedge$ OR	$\hat{m}_0, \hat{m}_h, r, \hat{r}', \hat{m}_t, \hat{b}, \hat{a}, \hat{r}_1, \hat{x} \| \hat{m}_0, \hat{m}_h, [\_], [\_], \hat{m}_t, \hat{b}, \hat{a}, \hat{r}_1, \hat{x}$	5	740	666
AND HA $\wedge$ NOT $\wedge$ OR	$\hat{b}, \hat{a}, \hat{r}', \hat{m}_t, \hat{r}, \hat{m}_0, \hat{r}', \hat{r}, \hat{r}_1, \hat{x} \| \hat{b}, \hat{a}, \hat{r}', \hat{m}_t, \hat{r}, \hat{m}_0, \hat{r}', [\_], \hat{r}_1, \hat{x}$	5	666	592

when combined and the AND proof can be executed with the optimizations discussed in Sect. 5. In the case of NOT $\wedge$ OR, both operators compute the $C, \tilde{C}$ commitments and only need to be obtained once. However, none of these operators enable the possibility of performing internal commitment reorganizations. Finally, AND $\wedge$ NOT $\wedge$ OR. This is the combination that enable us to perform a greater number of optimizations, First, $C, \tilde{C}, \hat{m}_t, \hat{r}$ do not need to be performed three times and the AND operator can be executed with internal commitment reorganization.

By performing external commitment reorganization we can obtain reductions in performance between 170.10 ms and 644.60 ms (Table 5) as well as in RAM (Table 6). This is mainly achieved where the three types of operators are being used and the commitments $C, \tilde{C}$ are reused together with the randomizers recomputed by the PRNG described in [18]. We rely on 5 attributes and on the cases created for the NOT operator (Sect. 5.2) together with the option where AND has the worst performance i.e. revealing all the attributes.

6   Conclusions

In this manuscript we have presented different strategies for implementing the operators for prime-encoded attributes described in [7]. We showed that when the number of attributes is large it can be possible to rely on prime-encoded proofs for improving the issuing process. Moreover, this also applies to the verification of a considerable amount of attributes. Besides, the selective disclosure operation can be improved in cases where hiding is needed by relying on prime-encoded attributes. Moreover, by externally and internally reordering the commitments involved in chained AND,OR and NOT operators it can be possible to obtain speed ups of 170.10-644.60 ms. These conclusions can be utilized as guidance in the creation of presentation policies when utilizing contemporary smart cards, taking into account that these operations are computational optimal in the target device in comparison to other implementation options.

References

1. Akhavi, A., Vallée, B.: Average Bit-Complexity of Euclidean Algorithms. In: Welzl, E., Montanari, U., Rolim, J.D.P. (eds.) ICALP 2000. LNCS, vol. 1853, pp. 373–387. Springer, Heidelberg (2000)
2. Akinyele, J.A., Garman, C., Miers, I., Pagano, M.W., Rushanan, M., Green, M., Rubin, A.D.: Charm: a framework for rapidly prototyping cryptosystems. J. Crypt. Eng. **3**(2), 111–128 (2013)
3. Bichsel, P., Camenisch, J., Groß, T., Shoup, V.: Anonymous credentials on a standard Java Card. In: ACM Conference on Computer and Communications Security, pp. 600–610 (2009)
4. Brands, S.A.: Rethinking Public Key Infrastructures and Digital Certificates: Building in Privacy. MIT Press, Cambridge (2000)
5. Camenisch, J., Dubovitskaya, M., Enderlein, R.R., Lehmann, A., Neven, G., Paquin, C., Preiss, F.-S.: Concepts and languages for privacy-preserving attribute-based authentication. J. Inf. Sec. Appl. **19**(1), 25–44 (2014)
6. Camenisch, J., Dubovitskaya, M., Lehmann, A., Neven, G., Paquin, C., Preiss, F.-S.: Concepts and languages for privacy-preserving attribute-based authentication. In: Fischer-Hübner, S., de Leeuw, E., Mitchell, C. (eds.) IDMAN 2013. IFIP AICT, vol. 396, pp. 34–52. Springer, Heidelberg (2013)
7. Camenisch, J., Groß, T.: Efficient attributes for anonymous credentials (extended version). IACR Cryptol. ePrint Arch. **2010**, 496 (2010)
8. Camenisch, J., Van Herreweghen, E.: Design and implementation of the idemix anonymous credential system. In: ACM Conference on Computer and Communications Security, pp. 21–30 (2002)
9. Camenisch, J.L., Lysyanskaya, A.: An efficient system for non-transferable anonymous credentials with optional anonymity revocation. In: Pfitzmann, B. (ed.) EUROCRYPT 2001. LNCS, vol. 2045, p. 93. Springer, Heidelberg (2001)
10. Camenisch, J.L., Lysyanskaya, A.: A signature scheme with efficient protocols. In: Cimato, S., Galdi, C., Persiano, G. (eds.) SCN 2002. LNCS, vol. 2576, pp. 268–289. Springer, Heidelberg (2003)
11. Camenisch, J.L., Stadler, M.A.: Efficient group signature schemes for large groups. In: Kaliski Jr., B.S. (ed.) CRYPTO 1997. LNCS, vol. 1294, pp. 410–424. Springer, Heidelberg (1997)
12. Chaum, D.: Security without identification: Transaction systems to make big brother obsolete. Commun. ACM **28**(10), 1030–1044 (1985)
13. Damgård, I.B.: Commitment schemes and zero-knowledge protocols. In: Damgård, I.B. (ed.) EEF School 1998. LNCS, vol. 1561, p. 63. Springer, Heidelberg (1999)
14. Damgård, I.B.: Efficient concurrent zero-knowledge in the auxiliary string model. In: Preneel, B. (ed.) EUROCRYPT 2000. LNCS, vol. 1807, pp. 418–430. Springer, Heidelberg (2000)
15. Fiat, A., Shamir, A.: How to prove yourself: practical solutions to identification and signature problems. In: Odlyzko, A.M. (ed.) CRYPTO 1986. LNCS, vol. 263, pp. 186–194. Springer, Heidelberg (1987)
16. Fujisaki, E., Okamoto, T.: Statistical zero knowledge protocols to prove modular polynomial relations. In: Kaliski Jr., B.S. (ed.) CRYPTO 1997. LNCS, vol. 1294, pp. 16–30. Springer, Heidelberg (1997)
17. Knuth, D.E.: The Art of Computer Programming, Volume II: Seminumerical Algorithms, vol. 2, 2nd edn. Addison-Wesley, Boston (1981)

18. de la Piedra, A., Hoepman, J.-H., Vullers, P.: Towards a full-featured implementation of attribute based credentials on smart cards. In: Gritzalis, D., Kiayias, A., Askoxylakis, I. (eds.) CANS 2014. LNCS, vol. 8813, pp. 270–289. Springer, Heidelberg (2014)
19. Sterckx, M., Gierlichs, B., Preneel, B., Verbauwhede, I.: Efficient implementation of anonymous credentials on java card smart cards. In: 1st IEEE International Workshop on Information Forensics and Security (WIFS), pp. 106–110. IEEE, London, UK, 2009 (2009)
20. Vullers, P., Alpár, G.: Efficient selective disclosure on smart cards using idemix. In: Fischer-Hübner, S., de Leeuw, E., Mitchell, C. (eds.) IDMAN 2013. IFIP AICT, vol. 396, pp. 53–67. Springer, Heidelberg (2013)

Software and System Security

Application of NTRU Using Group Rings to Partial Decryption Technique

Takanori Yasuda[1]($^{\boxtimes}$), Hiroaki Anada[1], and Kouichi Sakurai[1,2]

[1] Institute of Systems, Information Technologies and Nanotechnologies,
Fukuoka, Japan
{yasuda,anada,sakurai}@isit.or.jp
[2] Department of Informatics, Kyushu University, Fukuoka, Japan

Abstract. Partial decryption enables a ciphertext to be decrypted partially according to provided secret keys. In this paper, we propose a public key encryption scheme with the functionality of partial decryption. Our strategy is to use the NTRU cryptosystem. Under a design principle of the mathematical structure "group ring", we extend the original NTRU into group ring NTRU (GR-NTRU). First, we propose a generic framework of our GR-NTRU. Our GR-NTRU allows partial decryption with a single encryption process using a single public key. Besides, when we execute partial decryption under a secret key of GR-NTRU, we need no information to identify each part in a whole ciphertext. Consequently, management of a public key and a corresponding set of secret keys is rather easier than the naive method. Next, we propose a concrete instantiation of our generic GR-NTRU. A multivariate polynomial ring NTRU scheme is obtained by employing a product of different cyclic groups as the basis of the group ring structure. We will show examples of those new variants of NTRU schemes with concrete parameter values, and explain how we can employ them to use the functionality of partial decryption.

Keywords: NTRU · Lattice-based cryptography · Group ring · Partial decryption

1 Introduction

In a service network where many users participate in, encryption that enable a ciphertext to be decrypted by designated users are useful. Broadcast encryption [3,4] and attribute-based encryption [11,14] are well-known ones developed for the purpose in public-key framework. Partial decryption [2,10] is a notion for the kind of encryption. It enables a ciphertext to be decrypted *partially* according to provided secret keys. That is, when a plaintext is encrypted by a key, a user can recover only a part of the plaintext by using his secret key, while another

This research is commissioned by "Strategic Information and Communications R&D Promotion Programme (SCOPE), No. 0159-0016" Ministry of Internal Affairs and Communications, Japan.

M. Yung et al. (Eds.): INTRUST 2015, LNCS 9565, pp. 203–213, 2016.
DOI: 10.1007/978-3-319-31550-8_13

user can recover another part (or, a whole) of the plaintext by using another different secret key. A typical use-case of the partial decryption appears in the service of content delivery network for charged data such as movies and musics. In the service, partial disclosure attracts users but only users who purchased a whole-decryption key can enjoy content.

1.1 Previous Work

A naive method to realize partial decryption is to divide a plaintext into parts that correspond to receivers, and then encrypt each part by using each public key (in a public-key framework). A variant is that a content holder uses n public keys to encrypt n patterns of a part of a plaintext independently. Then a user, by using a subset of corresponding secret keys, decrypts those ciphertexts and obtains corresponding parts of the plaintext. Hence more (complete set of) secret keys can provide more parts of the plaintext (a whole of the plaintext, respectively). Bellare et al. [2] succeeded in achieving efficiency in this naive method by introducing a technique of re-usable randomness.

The naive method can also be applicable to common-key framework. One of the common-key methods, which is developed by Izu et al. [10], contains a masking procedure to hide some parts of a plaintext. Another variation is the hybrid one where a public-key encryption is applied to encrypt each common key (that is, each session key) for partial decryption.

A method of a different direction to achieve a similar goal is sanitizable signature [1,7]. In addition to ordinary functionalities of digital signature, it has a functionality that any part of a signed message can be santized keeping the validity of the legitimate signature (that is, it is not treated as falsification). Though it is different from partial decryption, it can be employed for the similar purpose to decode data partially.

We may imagine similarity and difference between partial decryption and secret sharing [13]. Santis et al. [12] proposed a scheme to share a function, which is called function sharing. In the scheme, an intractable function, that is, a hard-to-evaluate function is divided into shadow functions. Then, a subset of those shadow functions can collaborate together to effectively compute a value of the original function at a point only when the size of the subset is equal to or more than a threshold. Recently, Boyle et al. [6] proposed an extended notion called function secret sharing. If function sharing (FS) and function secret sharing (FSS) can be applied to a decryption function, the resulted functionality is similar to partial decryption, but FS and FSS are different in that where one cannot get any information when the number of shares is less than the threshold.

The naive method as well as the above related methods (except secret sharing) basically needs

- independent encryption processes by using independent keys,
- information to identify each part in a whole ciphertext,
- management of those independent keys and identifying information.

1.2   Our Contribution

In this paper, we propose a public key encryption scheme with the function-ality of partial decryption, which resolves the three problems. Our strategy is to use the NTRU cryptosystem [9], an algebraic lattice-based public-key cryp-tosystem. Under a design principle of the mathematical structure "group ring", we extend the original NTRU into group ring NTRU (GR-NTRU) to attain the functionality of partial decryption. As a first contribution, we propose a generic framework of our GR-NTRU to resolves the above three problems; actually, our GR-NTRU allows partial decryption with a single encryption process by using a single public key. Besides, when we execute partial decryption under a secret key of GR-NTRU, we need no information to identify each part in a whole ciphertext. If we use another different secret key, then we can decrypt another different part. Consequently, management of a public key and a corresponding set of secret keys is rather easier than the naive method.

As a second contribution, we propose two concrete versions of our generic GR-NTRU. Basically, the group ring structure allows us to apply finite groups as a basis of the group ring structure; if we apply a cyclic group, then the resulted scheme is the original NTRU [9]. If we apply a product of different cyclic groups, then we obtain a multivariate polynomial ring NTRU scheme. If we apply the Frobenius group, then we obtain a corresponding Frobenius NTRU scheme. We will show examples of those new variants of NTRU schemes with concrete parameter values, and explain how we can employ them to use the functionality of partial decryption.

This paper focuses on only application of GR-NTRU. Therefore, we do not discuss the security of GR-NTRU in this paper. Instead, another paper [15] analyzes the security of GR-NTRU whose authors include two authors of this paper.

2   NTRU

We review a description of the NTRU cryptosystem, not in the style of the original one [9], but in the style given in [8]. The NTRU scheme given in [8] has a more efficient algorithm of decryption than the original scheme.

Let N, p, q be integers satisfying $p < q$, and $R = \mathbb{Z}[x]/(x^N - 1)$. Any element f in R can be expressed uniquely as $f = \sum_{i=0}^{N-1} a_i x^i$ ($a_i \in \mathbb{Z}$). The subsets $\mathcal{L}_\mathbf{f}, \mathcal{L}_\mathbf{g}, \mathcal{L}_\mathbf{r}, \mathcal{L}_\mathbf{m}$ are defined as follows. First, we define the space of messages,

$$\mathcal{L}_\mathbf{m} = \left\{ f = \sum a_i x^i \in R \ \middle| \ -\frac{1}{2}(p-1) < a_i < \frac{1}{2}(p-1), \ \forall i \right\}. \qquad (1)$$

For positive integers d_1, d_2,

$$\mathcal{L}(d_1, d_2) = \left\{ f = \sum a_i x^i \in R \ \middle| \ \begin{array}{l} f \text{ has } d_1 \text{ coefficients equal } 1, \\ f \text{ has } d_2 \text{ coefficients equal } -1, \\ \text{the rest are } 0. \end{array} \right\}$$

For three integers $d_{\mathbf{f}}, d_{\mathbf{g}}, d$,

$$\mathcal{L}_{\mathbf{f}} = \mathcal{L}(d_{\mathbf{f}}, d_{\mathbf{f}} - 1), \ \mathcal{L}_{\mathbf{g}} = \mathcal{L}(d_{\mathbf{g}}, d_{\mathbf{g}}), \ \mathcal{L}_{\mathbf{r}} = \mathcal{L}(d, d). \tag{2}$$

Key Generation

Step 1 Choose $\mathbf{f} \in \mathcal{L}_{\mathbf{f}}$, $\mathbf{g} \in \mathcal{L}_{\mathbf{g}}$ such that there exists $F_q \in R$ satisfying $F_q *$
$(1 + p\mathbf{f}) = 1 \bmod q$.
Step 2 Let $\mathbf{h} = pF_q * \mathbf{g} \bmod q$.
Public Key $\mathbf{h}, p, q$.
Private Key $\mathbf{f}$.

Encryption. To encrypt a message $\mathbf{m} \in \mathcal{L}_{\mathbf{m}}$, we first choose randomly a $\mathbf{r} \in \mathcal{L}_{\mathbf{r}}$, then compute the ciphertext:

$$\mathbf{c} \equiv \mathbf{h} * \mathbf{r} + \mathbf{m} \bmod q.$$

Decryption. We compute

$$\mathbf{a} \equiv (1 + p\mathbf{f}) * \mathbf{c} \bmod q.$$

Here, we choose the coefficients of $\mathbf{a}$ in the interval from $-q/2$ to $q/2$. Then, $\mathbf{a}$ coincides with the message $\mathbf{m}$.

3   A Variant of NTRU Using Group Ring

In this section, we describe a NTRU-based cryptosystem using group ring, as an extension of NTRU.

3.1   Group Ring

Let G be a finite group.

Definition 1 [5]. $\mathbb{Z}[G]$ *is defined as the set*

$$\mathbb{Z}[G] = \left\{ \sum_{g \in G} a_g[g] \ \middle| \ a_g \in \mathbb{Z} \ (\forall g \in G) \right\}$$

Here, $[g]$ is a formal element associated to $g \in G$, and $\{[g] \,|\, g \in G\}$ becomes a basis of $\mathbb{Z}[G]$. The addition and multiplication in $\mathbb{Z}[G]$ are defined as follows:

(1) The addition is defined by component-wise addition.
(2) For any $g, h \in G$, $[g] * [h] = [gh]$. The multiplication of any two elements in $\mathbb{Z}[G]$ is defined by $\mathbb{Z}$-linear extension of the above formula.

By these addition and multiplication, $\mathbb{Z}[G]$ becomes a ring, which is called the group ring with respect to G.

Example 1. Let $C_N = \langle \sigma \rangle$ be a cyclic group of order N. Then $\mathbb{Z}[C_N]$ is isomorphic to $\mathbb{Z}[x]/(x^N - 1)$. In fact, the $\mathbb{Z}$-linear map below is a ring isomorphism.

$$\begin{array}{ccc} \mathbb{Z}[C_N] & \overset{\sim}{\longrightarrow} & \mathbb{Z}[x]/(x^N - 1) \\ \cup & & \cup \\ \sigma^i & \longmapsto & x^i. \end{array} \tag{3}$$

3.2 GR-NTRU

In order to apply NTRU-based cryptosystem using group ring to partial decryption technique widely, we prepare other sets of secret key, message etc. instead of $\mathcal{L}_\mathbf{f}, \mathcal{L}_\mathbf{g}, \mathcal{L}_\mathbf{r}, \mathcal{L}_\mathbf{m}$ in the original NTRU scheme. First, we define a L^∞-norm and L^2-norm of $f = \sum_{g \in G} a_g [g] \in R_G$ by

$$|f|_\infty = \max_{g \in G}\{|a_g|\}, |f|_2 = \left(\sum_{g \in G} |a_g|^2 \right)^{1/2},$$

respectively. Similarly as in [9], for any $\epsilon > 0$, there are $\gamma_1, \gamma_2 > 0$, depending on ϵ and G, such that the probability is greater than $1 - \epsilon$ that they satisfy

$$\gamma_1|F|_2|G|_2 \le |F * G|_\infty \le \gamma_2|F|_2|G|_2 \quad (F, G \in R_G).$$

For a positive number c, we define $\mathcal{M}(c) \subset R_G$ by

$$\mathcal{M}(c) = \{f \in R_G \,|\, |f|_2 < c\}.$$

Let p, q be integers satisfying $p < q$, G a finite group, and $R_G = \mathbb{Z}[G]$. Let $c_\mathbf{f}, c_\mathbf{g}, c_\mathbf{r}, c_\mathbf{m}$ be positive numbers satisfying

$$pc_\mathbf{g}c_\mathbf{r} < q/(4\gamma_2), \quad (1 + pc_\mathbf{f})c_\mathbf{m} < q/(4\gamma_2) \tag{4}$$

Key Generation

Step 1 Choose $\mathbf{f} \in \mathcal{M}(c_\mathbf{f}, \mathbf{g} \in \mathcal{M}_\mathbf{g}$ such that there exists $F_q \in R_G$ satisfying
$F_q * \mathbf{f} = 1 \bmod q$.
Step 2 Let $\mathbf{h} = F_q * \mathbf{g} \bmod q$.
Public Key $\mathbf{h}, p, q$.
Private Key $\mathbf{f}$.

Encryption. To encrypt a message $\mathbf{m} \in \mathcal{M}(c_\mathbf{m})$, we first choose a $\mathbf{r} \in \mathcal{M}(c_\mathbf{r})$, then compute the ciphertext:

$$\mathbf{c} \equiv \mathbf{h} * \mathbf{r} + \mathbf{m} \bmod q.$$

Decryption. We compute

$$\mathbf{a} \equiv (1 + p\mathbf{f}) * \mathbf{c} \bmod q.$$

Here, we choose the coefficients of $\mathbf{a}$ in the interval from $-q/2$ to $q/2$. Then, $\mathbf{a}$ coincides with the message $\mathbf{m}$.

We call this scheme *GR-NTRU* in this paper. We remark that in the case of $G = C_N$ (cyclic group), the corresponding GR-NTRU is equivalent to the original NTRU through the isomorphism (3).

3.3 Why Decryption Works

From (4), we have

$$|p\mathbf{g} \cdot \mathbf{r} + (1 + p\mathbf{f})m|_\infty < \frac{q}{2}. \tag{5}$$

Since

$$\mathbf{a} \equiv p\mathbf{g} \cdot \mathbf{r} + (1 + p\mathbf{f})\mathbf{m} \bmod q$$

and all the coefficient of $\mathbf{a}$ are chosen in the interval from $-q/2$ to $q/2$, $\mathbf{a}$ coincides with $p\mathbf{g} \cdot \mathbf{r} + (1 + p\mathbf{f})m$. Therefore, taking $\mathbf{a} \bmod p$, we can obtain $\mathbf{m}$.

4 Application to Partial-Decryption

Here, we explain how to apply GR-NTRU to partial-decryption method. What we need are two finite group G, H and a homomorphism between them, $\kappa :$ $G \to H$. From κ, we can construct a ring homomorphism between group rings, $\kappa_{\mathbb{Z}} : \mathbb{Z}[G] \to \mathbb{Z}[H]$ as follows

$$\kappa_{\mathbb{Z}}\left(\sum_{g \in G} c_g[g]\right) = \sum_{g \in G} c_g[\kappa(g)].$$

From two groups G, H, two GR-NTRU, NTRU_G and NTRU_H can be constructed. Then, a plaintext $\mathbf{m}_G$, ciphertext $\mathbf{c}_G$, secret key $\mathbf{f}_G$, public key $\mathbf{h}_G$ of NTRU_G are sent to $\mathbf{m}_H = \kappa_{\mathbb{Z}}(\mathbf{m}_G)$, $\mathbf{c}_H = \kappa_{\mathbb{Z}}(\mathbf{c}_G)$, $\mathbf{f}_H = \kappa_{\mathbb{Z}}(\mathbf{f}_G)$, $\mathbf{h}_H = \kappa_{\mathbb{Z}}(\mathbf{h}_G)$ via $\kappa_{\mathbb{Z}}$, respectively. If these satisfies the following conditions, $\mathbf{m}_H, \mathbf{c}_H, \mathbf{f}_H$ and $\mathbf{h}_H$ can be regarded as a plaintext, ciphertext, secret key, and public key of NTRU_H.

$$\mathbf{f}_H \in \mathcal{M}^H(c_{\mathbf{f}_H}), \ \mathbf{g}_H = \kappa_{\mathbb{Z}}(g_G) \in \mathcal{M}^H(c_{\mathbf{g}_H}), \ \mathbf{r}_H = \kappa_{\mathbb{Z}}(\mathbf{r}_G) \in \mathcal{M}^H(c_{\mathbf{r}_H}), \ \mathbf{m}_H \in \mathcal{M}^H(c_{\mathbf{m}_H}).$$

In what follows, we assume the above conditions, and apply the above homomorphism to partial-decryption. First, for a plaintext $\mathbf{m}_G$, $\mathbf{m}_H$ can be regarded as its partial information. Now, we consider the situation that Alice has an authority that the whole information $\mathbf{m}_G$ can be decrypted, and Bob has an authority that only partial information $\mathbf{m}_H$ can be decrypted. This situation can be realized as follows.

4.1   Key Generation

First, Alice generates a secret key $\mathbf{f}_G \in \mathcal{M}^G(c_{\mathbf{f}_G})$ and $\mathbf{g}_G \in \mathcal{M}^G(c_{\mathbf{g}_G})$. Next, Alice compute the corresponding public key $\mathbf{h}_G \in R_G$ from $\mathbf{f}_G$ and $\mathbf{g}_G$. $\mathbf{h}_G$ is the public key for Alice. The secret key for Bob is $\mathbf{f}_H := \phi_{\mathbb{Z}}(\mathbf{f}_G)$.

4.2   Encryption for Whole Plaintext

A plaintext $\mathbf{m}_G$ is identified with an element $\mathcal{M}^G(c_{\mathbf{m}_G})$. Here, we assume that $\phi_{\mathbb{Z}}(\mathbf{m}) \in \mathcal{M}^H(c_{\mathbf{m}_H})$ is satisfied. To encrypt $\mathbf{m}$, one chooses randomly $\mathbf{r}_G \in \mathcal{M}^G(c_{\mathbf{r}_G})$ such that $\phi_{\mathbb{Z}}(\mathbf{r}_G) \in \mathcal{M}^H(c_{\mathbf{r}_H})$. Next, one computes $\mathbf{e}_G \equiv \mathbf{h}_G \cdot \mathbf{r}_G + \mathbf{m} \bmod q$. $\mathbf{e}_G$ is the corresponding ciphertext.

4.3   Decryption for Alice

To decrypt $\mathbf{e}_G$, Alice computes $\mathbf{a}_G \equiv (1 + p\mathbf{f}_G) \cdot \mathbf{e}_G \bmod q$. Here, Alice chooses all the coefficients of $\mathbf{a}_G$ in the interval from $-q/2$ to $q/2$. Then, $\mathbf{a}_G$ coincides with the message $\mathbf{m}_G$.

4.4   Decryption for Bob

First, Bob computes $\mathbf{e}_H = \phi_{\mathbb{Z}}(\mathbf{e}_G)$. Next, Bob computes $\mathbf{a}_H \equiv (1 + p\mathbf{f}_{II}) \cdot \mathbf{e}_H \bmod q$. Here, Bob chooses all the coefficients of $\mathbf{a}_H$ in the interval from $-q/2$ to $q/2$. Then, $\mathbf{a}_H$ coincides with the message $\mathbf{m}_H$.

5   Application Using Multivariate NTRU

NTRU employs the group ring associated with the cyclic group C_N. If we change this group to a product of cyclic groups $C_{N_1} \times \cdots \times C_{N_l}$, then we have the multivariate NTRU. Let us consider the case that $l = 3$ and N_1, N_2, N_3 are all prime. Let p, q be positive integers. Then we have the following ring isomorphism:

$$R_3 = \mathbb{Z}[x_1, x_2, x_3]/(x_1^{N_1} - 1, x_2^{N_2} - 1, x_3^{N_3} - 1)$$
$$\simeq \mathbb{Z}[C_{N_1} \times C_{N_2} \times C_{N_3}]$$

In the case of $l = 2$, we have three associated rings:

$$R_2^1 = \mathbb{Z}[x_2, x_3]/(x_2^{N_2} - 1, x_3^{N_3} - 1) \simeq \mathbb{Z}[C_{N_2} \times C_{N_3}],$$
$$R_2^2 = \mathbb{Z}[x_1, x_3]/(x_1^{N_1} - 1, x_3^{N_3} - 1) \simeq \mathbb{Z}[C_{N_1} \times C_{N_3}],$$
$$R_2^3 = \mathbb{Z}[x_1, x_2]/(x_1^{N_1} - 1, x_2^{N_2} - 1) \simeq \mathbb{Z}[C_{N_1} \times C_{N_2}]$$

Any element f_3 in the ring R_3 is expressed uniquely by

$$f_3 = \sum_{i_1, i_2, i_3} a_{i_1, i_2, i_3} x_1^{i_1} x_2^{i_2} x_3^{i_3} \quad (a_{i_1, i_2, i_3} \in \mathbb{Z}).$$

In the case of two variables, for example, an element f_2 in R_2^3 is expressed by the form, $f_2 = \sum_{i_1,i_2} a'_{i_1,i_2} x_1^{i_1} x_2^{i_2}$ ($a'_{i_1,i_2} \in \mathbb{Z}$). Considering the natural group homomorphism,

$$\phi_1 : C_{N_1} \times C_{N_2} \times C_{N_3} \to C_{N_2} \times C_{N_3},$$
$$\phi_2 : C_{N_1} \times C_{N_2} \times C_{N_3} \to C_{N_1} \times C_{N_3},$$
$$\phi_3 : C_{N_1} \times C_{N_2} \times C_{N_3} \to C_{N_1} \times C_{N_2},$$

the corresponding ring homomorphisms are described as follows:

$$\phi_{1,\mathbb{Z}}^1 : R_3 \ni f(x_1, x_2, x_3) \mapsto f(1, x_2, x_3) \in R_2^1,$$
$$\phi_{2,\mathbb{Z}}^1 : R_3 \ni f(x_1, x_2, x_3) \mapsto f(x_1, 1, x_3) \in R_2^2,$$
$$\phi_{3,\mathbb{Z}}^1 : R_3 \ni f(x_1, x_2, x_3) \mapsto f(x_1, x_2, 1) \in R_2^3.$$

Let a plaintext $\mathbf{m} \in R_3$ be given by

$$\mathbf{m} = \sum_{i_1,i_2,i_3} a_{i_1,i_2,i_3} x_1^{i_1} x_2^{i_2} x_3^{i_3} \quad (a_{i_1,i_2,i_3} \in \mathbb{Z}).$$

Then the following three partial informations $\mathbf{m}_1 \in R_2^1, \mathbf{m}_2 \in R_2^2, \mathbf{m}_3 \in R_2^3$ of $\mathbf{m}$ are considered:

$$\mathbf{m}_1 = \phi_{1,\mathbb{Z}}^1(\mathbf{m}) = \sum_{i_2,i_3} (\sum_{i_1} a_{i_1,i_2,i_3}) x_2^{i_2} x_3^{i_3},$$
$$\mathbf{m}_2 = \phi_{2,\mathbb{Z}}^1(\mathbf{m}) = \sum_{i_1,i_2} (\sum_{i_2} a_{i_1,i_2,i_3}) x_1^{i_1} x_3^{i_3},$$
$$\mathbf{m}_3 = \phi_{3,\mathbb{Z}}^1(\mathbf{m}) = \sum_{i_1,i_2} (\sum_{i_3} a_{i_1,i_2,i_3}) x_1^{i_1} x_2^{i_2}.$$

6 Combination of Several Partial Informations

In this section, we introduce a method to decrypt several partial informations of different types as an application of several homomorphisms.

6.1 Application 1

Let $\phi_1, \phi_2 : R_3 \to R_2^3$ be defined by

$$\phi_1(f(x_1, x_2, x_3)) = f(x_1, x_2, 1),$$
$$\phi_2(f(x_1, x_2, x_3)) = f(x_2, x_1, 1).$$

For a plaintext $\mathbf{m} \in R_3$, $\mathbf{m}_1 = \phi_1(\mathbf{m})$, $\mathbf{m}_2 = \phi_2(\mathbf{m})$ are regarded as partial informations of $\mathbf{m}$. Figure 1 shows that A can decrypt whole information $\mathbf{m}$ and B_1, B_2 can decrypt partial information m_1, m_2, respectively. The top node of this graph has two child nodes. Furthermore, we can consider a graph with a lot of siblings.

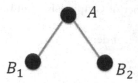

Fig. 1. Decryption of two different partial informations

6.2 Application 2

Let $\phi_1 : R_3 \to R_2^3$, $\phi_3 : R_2^3 \to R$ be defined by

$$\phi_2(f(x_1, x_2, x_3)) = f(x_1, x_2, 1),$$
$$\phi_3(f(x_1, x_2)) = f(x, 1).$$

For a plaintext $\mathbf{m} \in R_3$, $\mathbf{m}_2 = \phi_2(\mathbf{m})$ is a partial information of $\mathbf{m}$ and $\mathbf{m}_3 = \phi_3(\mathbf{m}_2)$ is a partial information of $\mathbf{m}_2$. Figure 2 shows that A can decrypt whole information $\mathbf{m}$ and B_2, C_1 can decrypt partial information $\mathbf{m}_2, \mathbf{m}_3$, respectively. The graph expresses that upper node is nearer whole information. If we assume $N_1 = N_2 = N_3 = N$, instead of ϕ_3, we can use

$$\phi_3'(f(x_1, x_2)) = f(x, x^i) \quad (i = 1, \ldots, N-1),$$

therefore, there exist more than N homomorphisms which corresponds to the same graph as above (Table 1).

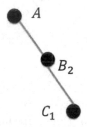

Fig. 2. Decryption of two partial information of containment relationship

6.3 Application 3

Combining Application 1 and 2, the following partial decryption scheme can be realize: Let $\phi_1, \phi_2 : R_3 \to R_2^3$, $\phi_3, \phi_4 : R_2^3 \to R$ be defined by

$$\phi_1(f(x_1, x_2, x_3)) = f(x_1, x_2, 1),$$
$$\phi_2(f(x_1, x_2, x_3)) = f(x_2, x_1, 1),$$
$$\phi_3(f(x_1, x_2)) = f(x, 1),$$
$$\phi_4(f(x_1, x_2)) = f(x, x).$$

Table 1. Number of Graph using multivariate NTRU

n	♯ of depth 1	♯ of depth 2
3	3	$3N$
4	4	$4N^2$
5	5	$5N^3$
6	6	$6N^4$
7	7	$7N^5$
8	8	$8N^6$
9	9	$9N^7$
10	10	$10N^8$

Then a partial decryption scheme corresponding to the graph of Fig. 3 can be constructed. In the case of the multivariate NTRU using 3 variables, there appears no graph of depth more than 2, but in general, in the case of the multivariate NTRU using n variables, the graph of depth $n-1$ appears.

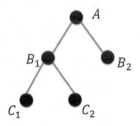

Fig. 3. Decryption of 4 partial informations

7 Concluding Remark

We propose a method to realize several type of partial decryption using lattice-based cryptography using group rings. In fact, diversity of several groups and homomorphisms between them provides several types of partial decryption schemes. As a non-trivial example, we present an application using the multivariate NTRU.

Given several finite groups and several homomorphisms between corresponding group rings, we can design a partial decryption system. We will investigate how many combination of finite groups and homomorphisms can be applied to partial decryption system as a future work.

References

1. Ateniese, G., Chou, D.H., de Medeiros, B., Tsudik, G.: Sanitizable signatures. In: di Vimercati, S.C., Syverson, P.F., Gollmann, D. (eds.) ESORICS 2005. LNCS, vol. 3679, pp. 159–177. Springer, Heidelberg (2005)

2. Bellare, M., Boldyreva, A., Staddon, J.: Randomness re-use in multi-recipient encryption schemeas. In: Desmedt, G. (ed.) PKC 2003. LNCS, vol. 2567, pp. 85–99. Springer, Heidelberg (2003)
3. Berkovits, S.: How to broadcast a secret. In: Davies, D.W. (ed.) EUROCRYPT 1991. LNCS, vol. 547, pp. 535–541. Springer, Heidelberg (1991)
4. Boneh, D., Gentry, C., Waters, B.: Collusion resistant broadcast encryption with short ciphertexts and private keys. In: Shoup, V. (ed.) CRYPTO 2005. LNCS, vol. 3621, pp. 258–275. Springer, Heidelberg (2005)
5. Bovdi, A.A.: Group Algebra. Springer Publishing Company, Incorporated (2001)
6. Boyle, E., Gilboa, N., Ishai, Y.: Function secret sharing. In: Oswald, E., Fischlin, M. (eds.) EUROCRYPT 2015. LNCS, vol. 9057, pp. 337–367. Springer, Heidelberg (2015)
7. Brzuska, C., Fischlin, M., Lehmann, A., Schröder, D.: Santizable signatures: how to partially delegate control for authenticated data. In: Proceedings of the Special Interest Group on Biometrics and Electronic Signatures BIOSIG 2009, 17-18 September 2009 in Darmstadt, Germany, pp. 117–128 (2009)
8. Hoffstein, J., Howgrave-Graham, N., Pipher, J., Silverman, J., Whyte, W.: Hybrid lattice reduction and meet in the middle resistant parameter selection for ntruencrypt
9. Hoffstein, J., Pipher, J., Silverman, J.H.: NTRU: a ring-based public key cryptosystem. In: Buhler, J.P. (ed.) ANTS 1998. LNCS, vol. 1423, pp. 267–288. Springer, Heidelberg (1998)
10. Izu, T., Ito, K., Tsuda, H., Abiru, K., Ogura, T.: Privacy-protection technologies for secure utilization of sensor data. Fujitsu Sci. Tech. J. $50(1)$, 30–33 (2014)
11. Sahai, A., Waters, B.: Fuzzy identity-based encryption. In: Cramer, R. (ed.) EUROCRYPT 2005. LNCS, vol. 3494, pp. 457–473. Springer, Heidelberg (2005)
12. Santis, A.D., Desmedt, Y., Frankel, Y., Yung, M.: How to share a function securely. In: Proceedings of the Twenty-Sixth Annual ACM Symposium on Theory of Computing, 23–25 May 1994, Montréal, Québec, Canada, pp. 522–533 (1994)
13. Shamir, A.: How to share a secret. Commun. ACM $22(11)$, 612–613 (1979)
14. Yamada, S., Attrapadung, N., Santoso, B., Schuldt, J.C.N., Hanaoka, G., Kunihiro, N.: Verifiable predicate encryption and applications to CCA security and anonymous predicate authentication. In: Proceedings 15th International Conference on Practice and Theory in Public Key Cryptography PKC–2012, Darmstadt, Germany, May 21–23 2012, pp. 243–261 (2012)
15. Yasuda, T., Dahan, X., Sakurai, K.: Characterizing NTRU-variants using group ring and evaluating their lattice security. To be appear as an IACR e-print paper

RbacIP: A RBAC-Based Method for Intercepting and Processing Malicious Applications in Android Platform

Li Lin[1,2,3], Jian Ni[1,2]([✉]), Jian Hu[1,2], and Jianbiao Zhang[1,2,3]

[1] College of Computer Science, Beijing University of Technology, Beijing, China
18733192581@163.com
[2] Beijing Key Laboratory of Trusted Computing, Beijing, China
[3] National Engineering Laboratory for Critical Technologies of Information Security Classified Protection, Beijing, China

Abstract. With the rapid development of Android-based smart phones and pads, android applications show explosive growth. Because third-party application market regulation is lax, many normal applications are embedded malicious code and then many security issues occur. The existing antivirus software cannot intercept malicious behaviors from those repackaged applications in many cases. To solve these problems, we propose a new method called *RbacIP*, which integrates RBAC into intercept and disposal process of malicious android applications. In *RbacIP*, the malicious behaviors of applications are monitored by inserting Linux kernel function call dynamically. Exploiting the Netlike technology, the information of malicious behaviors are feedback from the kernel layer to the user layer. On the user layer, depending on the roles assigned, android applications are authorized to the corresponding permissions. According to the characteristics of RBAC, it can achieve the minimum authorization for malicious applications. Meanwhile, to balance the user experience and his privacy protection needs, users are allowed to make fine-grained decision based on RBAC policy, rather than permit or prohibit. Finally, we implemented RbacIP in real android platform. Comprehensive experiments have been conducted, which demonstrate the effectiveness of the proposed method by the comparison with traditional HIPS systems at the malicious programs detection performance and resource consumption.

Keywords: RBAC · Hook · Dynamic detection · Android-ndk

1 Introduction

The current mainstream malware detection method is divided into static detection and dynamic monitoring, of which the main ways in the static detection include: analyzing applications signature information, APK source code reverse analysis, malicious code library matching. Static detection has higher malicious

© Springer International Publishing Switzerland 2016
M. Yung et al. (Eds.): INTRUST 2015, LNCS 9565, pp. 214–225, 2016.
DOI: 10.1007/978-3-319-31550-8_14

code library dependencies, when there is a new type of malicious program, which often cannot accurately identify malicious behavior, so the detection rate and accuracy of malicious programs should be improved. Dynamic monitoring by actually running the application, which is expected to trigger malicious behavior, to achieve the purpose of detecting malicious programs. The feature of classic dynamic detection methods is with little dependency for the special code repository, by capturing the real running of malicious acts to determine whether it was malicious software. There are the risk of malicious programs bypassing the system detects.

To solve these problems, this paper proposes *RbacIP*: a RBAC [7]-based Method for Intercepting and Processing Malicious Applications in Android Platform. Compared with other research work, this dissertations innovation points are as follows:

(1) This method proposes *RbacIP*: a RBAC-based Method for Intercepting and Processing Malicious Applications in Android Platform, By defining the different roles of least privilege portfolio to achieve the malicious programs minimum authorized, which could detect and dispose the application steal user privacy malicious behavior.
(2) Depending on different APP application scenarios, when a malicious accessing user privacy data, the method will according the APPs permissions information corresponding to the role, deciding whether to allow applications related operations, giving users more accurate and granular security tips, while allowing the user to choose by themselves.
(3) The method by modifying the Android Linux kernel to implement malicious behavior dynamic detection, because the application cannot modify the kernel layer of function calls, thus avoiding the risk of malicious programs to bypass the traditional HIPS system detects, with greater security.

The rest of the paper is organized as follows: Sect. 2 describes related work; Sect. 3 describes *RbacIP* in detail; Sect. 4 is the experimental results and analysis; and finally we concludes the paper and proposes future research work.

2   Related Work

Android malware detection has been the hot topic in recent years, the current mainstream malware detection method is divided into two categories: static test or dynamic test.

Static detections main way is to analyze the permission information of the application, comparing the signature information of the application, through reverse engineering to analyze the source of the APK and so on. Stephen et al. [2] By analyzing Manifest.xml file additional information in addition to the permissions, they proposed Manilyzer, the system can take advantage of the wealth of information of the manifest.xml automatically generated feature vectors, while it use modern machine learning algorithm for applications testing classification. Patrick et al. [3] proposed a feature set that contains the list of android malware permissions and collection of API calls, classification across use the permissions of feature set contained and the characteristic information of API to achieve the detection of malicious software.

The major way of dynamic detection adapts to dynamically run applications, by triggering its malicious behavior to achieve the detection of malicious programs. Commonly used dynamic testing tools include Monkey, MonkeyRunner, TaintDroid and droidBox and so on. Mingshen et al. [5] designed and implemented a secure, scalable HIPS (Host-based Intrusion Prevention System) platform - "Patronus". "Patronus" not only provides intrusion prevention while eliminating the need to modify the Android system, it can also dynamically detect existing malware based on Android runtime information. Experiments show, "Patronus" can effectively prevent invasive behavior and accurately detect malware, with a very low performance overhead and power consumption.

3   Problem Statement

As shown in Fig. 1, when users install applications in Android device, Android system will list about the application applied for the privileges information and these rights may result in the risk of information leakage of user privacy, the user clicks on the consent, then the APP will be installed on the Android device, otherwise APP will not be installed. Android system authority information reaches as many as 151, when Android applications access user privacy data, they must have the appropriate permissions information to operate. But most users did not notice the rights information Android APP applied when installing APP, which gives an opportunity to be exploited by malicious applications, indirectly pose a threat to users' privacy security. This paper is conducted for such the kinds of scenarios.

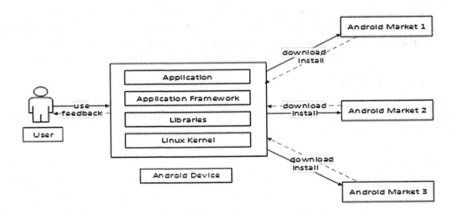

Fig. 1. Application scenario

4   Design of RbacIP

Android system has classic fourtier structure model, from top to bottom in turn is the Application layer, Application Framework layer, Library layer, and Linux kernel layer. Any operation of the application all need to call the function of the Linux kernel. The *RbacIP* method include four modules, malicious behavior detection module, communication module, intercepting prompt module and RBAC-based intercept module. The malicious behavior detection module is deployed to achieve the key functions of the system Hook calls based on Linux Hook technology. Communication module passes the application malicious behavior found in malicious behavior detection module to intercept prompt module for further disposal. Based on JNI and android-ndk technology, interception prompt module deployed in user layer gives the corresponding prompt to the user. When an application requests access to the user's private data, RBAC-based interceptor module decide whether allow the application related operations by querying RBAC policy library. The framework of *RbacIP* is showed in Fig. 2.

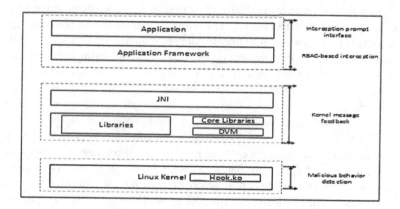

Fig. 2. The framework of RbacIP

4.1   Malicious Behavior Detection

Linux kernel layer is divided into kernel mode and user mode, the system call interface located between the core mode and user mode, it encapsulates underlying operating to provide the interface the functions can be invoked to the upper layer. Such as user mode program request read operation to kernel mode program, after the operating system obtains control returns the results to the user. Linux systems generally process is unable to read the system kernel, the system call is the only entrance into the core mode. Linux kernel using system call number only represents a system call, when accessing system calls the process needs query from sys_call_table (system call table) addressing to achieve. When a user program execution system call will first execute int 0x80 software interrupt instruction, which makes Linux execution transition from user mode to kernel mode, while calling sys_call_table to call the corresponding system function. Because Android applications will eventually call the Linux system function, so we can pass Hook of Linux system function calls, to achieve the purpose of monitoring malicious behavior of the application. Linux Hook is a replacement for the Linux system calls function, before the application requests a system call, we will point the corresponding system call table to own definition of the calling function, after the appropriate action and then return to normal the system call. The method implements monitoring an application malicious acts, including reading the user contacts, location information, communications records, text messages and other private data, while across communication module will transfer malicious acts to intercept prompt module accordingly make interception and disposal.

4.2   Kernel Message Feedback

We use netlink socket to complete the communication kernel mode and user mode. Kernel mode applications through function netlink_kernel_create() to initialize netlink socket connection. At this time, the user mode program through function socket() to create a user mode socket, which needs to specify the user mode socket's address field, protocol type. And for establishing a communication connection with kernel mode programs, kernel mode and user mode application need to use the same protocol type. User mode application by bind() function to achieve relevance each other between source socket address and open socket address. Linux transmit messages from user mode to kernel mode, by function sendmsg(fd, &msg, 0) to achieve it, through the function recvmsg(fd, &msg, 0) to accept messages from the kernel. User mode program needs to send its own address, the process id and other fields to kernel mode program, when the kernel mode program to monitor the application of reading user privacy records and other acts, which will according to process id and address from the user mode to send a message to specify the location.When user mode application program receives the message from the kernel mode, it will give user prompts by the way of graphical, while according to the user's selection return the corresponding results back to the kernel mode program. When the kernel mode application program receives the message from the user mode, it will invoke hello_nl_recv_msg() to process the message from user mode, according to the user's choice to decide whether to allow the application of related operations.

4.3   Interception Prompt Interface

To provide the user-friendly interface, we use *JNI* technology to achieve giving the user fine-grained system prompts in the form of pop-up box. We achieve *JAVA* code and *C* code call each other based Android.mk, *C* programs and .h header files. Wherein, Android.mk is similar to the Makefile file in Linux system, which defines the rules for compiling .*C* files, only add the file can we through "ndk-build" script to compile C programs. Jni file directory .h header are generated by javah command through .java program corresponding header files, javah command will generate .java program using native keyword modification method into methods signatures can be referenced by .C file. Jni directory .C Files is to be compiled C program, we need to introduce .h header file which was generated by the command javah, and then implement the relevant functions defined in function.The method is based .C file to achieve user-level socket initialization tasks, achieve receiving kernel mode messages and call related $JAVA$ program in a manner of graphical interface to the user. C program in the internal of $JNIENV$ * env contains a pointer to the Java VM Dalvik virtual machine object, whenever a Java thread firstly call C program, Dalvik virtual machine all will produce a $JNIEnv$ * pointer instances for the Java thread.

4.4 RBAC-Based Interception

The method is for to protect the user's privacy, which introduce RBAC-based access control mechanism. In this way it is not the application was installed then it can access all of the resources within its competence allowed range, when the application access user privacy data, this method will query the application belongs roles and rights information corresponding to the role. The system will query RBAC policy repository, to decide whether to allow the application of the corresponding request. At the same time, this approach also supports the users to choose by themselves, in case the user has not selected the system will default to perform operations according to the query result of the RBAC policy library. Thus the method meet user needs, while protecting the user's privacy.

Set R $(R_1, R_2, ..., R_n)$ represents the definition of the roles, the method according to the functional requirements of existing daily Android applications, which defined Android applications defined as six roles, roles are defined as: $R_1, R_2, R_3, R_4, R_5, R_6$, which represent: audio-visual, map, communications, payment type, shopping class and games. The set P $(P_1, P_2, ..., P_n)$ means that the definition of privileges, the method summarized likely to leak user privacy 20 right information, whose permissions are defined as follows: $P_1, P_2, P_3, ..., P_{20}$, which are $ACCESS\_COARSE\_LOCATION,$ $ACCESS\_FINE\_LOCATION,$ $WRITE\_SMS,$ $WRITE\_OWNER\_DATA,$ $WRITE\_CONTACTS,$ $SEND\_$ $SMS,$ $RECORD\_AUDIO,$ $RECEIVE\_SMS,$ $RECEIVE\_MMS,$ $READ\_PHONE\_SMS,$ $READ\_OWNER\_DATA, READ\_CONTACTS, PROCESS\_OUTGOING\_CALLS,$ $INTERNET, CHANGE\_WIFI\_STATE, CHANGE\_NETWORK\_STATE, CALL\_$ $PHONE, BROADCAST\_SMS, ACCESS\_ WIFI\_ STATE, ACCESS\_NETWORK\_$ $STATE.$

Function G $(R,\ P)$ represents the definition of relationship assignment between roles and permissions, specifying the roles and corresponding rights information, such as G $(R_m, P_n) = 1$ indicates role R_m has permissions P_n, G $(R_o, P_w) = 0$ indicates Role R_o does not have permission P_w. To protect users' privacy and security, this method assigned to the role for the corresponding minimum permissions collection which was required to complete its task. Such as Communications applications function is to add contacts friends, send instant messages, send text messages, etc., completing its tasks permissions must include *SEND SMS, RECEIVE SMS, READ PHONE SMS, READ CON-TACTS, INTERNET, CALL PHONE* six right information, in order to protect the user's privacy security, the method assigned communications APP role only for the minimum permissions to complete the tasks set. When communications APP requesting user privacy data privilege beyond its minimum permissions set, the system default reject the application request. Function $W(R_1, R_2)$ represent mutually exclusive roles relationship definition, the application cannot be allocated two mutually exclusive roles, such as $W(R_m, R_n) = 1$ indicates role R_m and R_n are not mutually exclusive roles, applications can have the role R_m and R_n, $W(R_s, R_v) = 0$ indicates roles R_s and R_v are two mutually exclusive roles, the application cannot have both roles R_s and R_v. This method in order to protect user privacy and security, which defined the six roles exclusive relationship. For example, the role of communication and payment Role belong to exclusive role, when the application is assigned communication role, which cannot be allocated to pay class role. Specific roles mutually exclusive relationship is defined in Table 1.

Table 1. Exclusive role relationship definition table

	Video	Map	Communication	Payment	Shop	Game
Video	1	0	0	0	0	1
Map	0	1	0	0	0	0
Communication	0	0	1	0	0	0
Payment	0	0	0	1	0	0
Shop	0	0	0	0	1	0
Game	1	0	0	0	0	1

Function $R_a(P_1, P_2, ..., P_n)$ represents the permissions definition set of role R_a, function $R_{a+1}(P_1, P_2, ..., P_n, P_{n+1})$ represents the privileges definition set of the role R_{a+1}, R_{a+1} represent R_a new privileges P_{n+1} for the role, the system dynamic update RBAC policy library after new added roles information. The method is to balance the needs of the user, RBAC policy library that supports dynamic updates. In the initial installation of the application is running, the system will assign the appropriate roles according to the application's functionality, the role includes the minimum permissions set for the application to complete its task. When an application requests permission beyond its minimum permissions set, the system will give the user the relevant safety tips. When users agree to grant the application relative authority, the system will assign a new role to application according to user selection. The new role contains the permissions was requested by the application of beyond the minimum permissions set. When the application requests the permission next time, the system will automatically allow or deny the application related operations according the previous request record.

4.5   Interception and Disposal Algorithm

The interception and disposal algorithm is shown in Table 2.

Table 2. Interception and disposal algorithm in *RbacIP*

Input: application to be tested
Output: RBAC-based system

Begin:
1. Install android app into android devices
2. Run the app, when the app visit users privace data, query RBAC
strategy database.
3. According apps role and the permission of the role to decide
whether approve apps request.
4. If(the roles permission status == 1){
  Pop window to suggest user approve apps request.
    If(user click approve){
      Feedback to kernel, load app request resource
      }
    Else{
      Feedback to kernel, refuse app request resource
      }
  }
5. Else{
  Pop window to suggest user refuse apps request.
    If(user click approve){
      Feedback to kernel, load app request resource
      }
    Else{
      Feedback to kernel, refuse app request resource
      }
  }
End

---

5   Performance Evaluation

This section tests the effectiveness of *RbacIP* method in real Android platform, through comparing with the current prevalent HIPS method.

5.1   Experimental Environment

We have tested resource consumption of the method and interception result to android system, across inserting the compiled kernel module into the Android Linux kernel, which realized monitoring application malicious behavior in the Android Linux kernel layer. When detecting the application reads user privacy data, the system will base on the roles of the APP, and the corresponding permissions to give users more fine-grained, more accurate tips. The user can across the system prompts make their own choice, in case the user does not make a choice, the program will perform system operations, the last kernel layer according to the user choice to decide whether to allow or prohibit specific malicious APP's operations.

5.2   Interception Result Analysis

We have selected the TOP 100 applications from millet application store as normal samples, and downloaded some malicious applications from third-party security testing center, in order to test the method, we have written some applications for testing, a total of 100 malicious programs Sample. We have tested the *RbacIP* method and the traditional HIPS method. When application reading user privacy, the *RbacIP* method will prompt the user in the form of bomb box with malicious behavior. System bombs box example is shown in Fig. 3.

Fig. 3. An example of program sensitive operations intercepted

We have tested programs during they reading the user address book, SMS record, backstage networking and other acts, whether the system gives the user about the corresponding prompt according to RBAC policy library. Though testing, the method for sensitive operation of the application, which can give the appropriate prompt, in case the user does not operate, it will default execute the corresponding operation according to the system prompts. In particular, when the application reads user privacy data, and the method have a high detection rate for these behaviors, specifically shown in Fig. 4, the figure A-G successively representing access location information, access communication records, read contacts, read SMS records, send text messages, read e-mail messages, connect to the network. Meanwhile, as shown in Fig. 5, when the malicious application access the user's privacy records, the *RbacIP* method has higher system interception rate by comparison with the traditional HIPS.

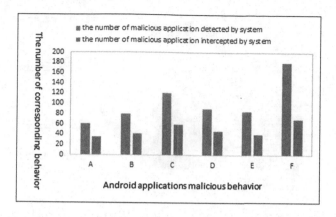

Fig. 4. Different malicious behaviors of Android application vs the number of malicious application detected and intercepted by system under *RbacIP*

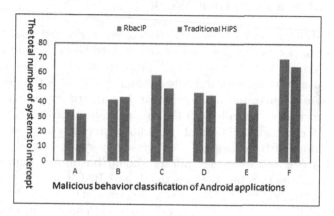

Fig. 5. Different malicious behaviors of Android application vs the number of malicious application intercepted under *RbacIP* and traditions HIPS

5.3   Resource Consumption Analysis

The method has been tested by Android system, Android kernel version uses android 4.4, the screen resolution is 1280*760, Android corresponding API version is 19, the memory size is 1G. By inserting Hook kernel modules in the Linux kernel layer to achieve malicious behavior detection, so the Android device's memory consumption is almost zero. At the application layer, by installing APP to achieve bomb box functionality, through testing the memory of the APP is less than 1 %. After testing, this method of phone system resource consumption is low. Specific consumption of system resources shown in Fig. 6.

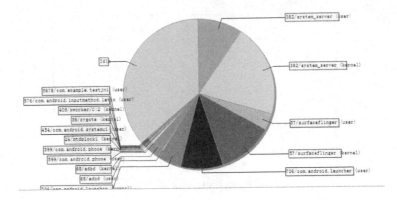

Fig. 6. Resource consumption ratio of *RbacIP* in real Android platform

6   Conclusion and Future Work

When users install Android applications, who often overlooked the APP applied authority information, which gives an opportunity to be exploited by malicious applications. Aiming at this situation, we propose *RbacIP*. When a user inadvertently install malicious software, this method can achieve the detection and interception of malicious application behavior. In order to protect users' privacy, assigning the corresponding role with the minimum permissions to complete its mandate. It will cause some of normal APPs not to access some resources, which served as malicious theft behavior of user privacy by system mistake. Therefore, RBAC-based access control policy library should be further improved, using big data and machine learning methods to adjust RBAC policy database intelligently. In the future, we plan to improve the detection rate of malicious programs by adjusting the RBAC policy repository dynamically based on machine learning methods.

Acknowledgement. This work is supported by grants from the China National Science Foundation (Project No. 61502017), China 863 High-tech Programme (Project No. 2015AA016002). The authors would like to thank the anonymous reviewers for their constructive comments.

References

1. Alazab, M., Moonsamy, V., Batten, L., et al.: Analysis of malicious and benign android applications. In: 2012 32nd International Conference on Distributed Computing System Workshops, pp. 608–616 (2012)
2. Stephen, F., Dillon, S., Bing, W.: Manilyzer: automated android malware detection through manifest analysis. In: IEEE 11th International Conference on Mobile Ad Hoc and Sensor Systems, pp. 767–772 (2014)
3. Patrick, P., Wen-Kai, S.: Static detection of android malware by using permissions and API calls. In: International Conference on Machine Learning and Cybernetics, pp. 82–87 (2015)

4. Daiyong, Q., Lidong, Z., Fan, Y., et al.: Detection of android malicious apps based on the sensitive behaviors. In: IEEE 13th International Conference on Trust, Security and Privacy in Computing and Communications, TrustCom, pp. 877–883 (2014)
5. Mingshen, S., Min, Z., John, C., et al.: Design and implementation of an android host-based intrusion prevention system. In: Proceedings of the 30th Annual Computer Security Applications Conference (2014)
6. Wen-Chieh, W., Shih-Hao, H.: DroidDolphin: a dynamic android malware detection framework using big data and machine learning. In: Proceedings of the 2014 Conference on Research in Adaptive and Convergent Systems (2014)
7. Qiang, W., Jason, C., Konstantin, B., et al.: Authorization recycling in hierarchical RBAC systems. ACM Trans. Inf. Syst. Secur. **14**, 3 (2011)
8. Reinhard, T., Julio, S., Wolfgang, S., et al.: Dead or alive: finding zombie features in the linux kernel. In: Proceedings of the First International Workshop on Feature-Oriented Software Development (2009)
9. Jemin, L., Hyungshin, K.: Framework for automated power estimation of android applications. In: Proceeding of the 11th Annual International Conference on Mobile Systems, Applications, and Services (2013)
10. Zheng, S., Shijia, P., Yu-Chi, S., et al.: Headio: zero-configured heading acquisition for indoor mobile devices through multimodal context sensing. In: Proceedings of the 2013 ACM International Joint Conference on Pervasive and Ubiquitous Computing (2013)
11. Sebastian, F., Bert, A., Gerhard, T., et al.: CoenoFire: monitoring performance indicators of firefighters in real-world missions using smartphones. In: Proceedings of the 2013 ACM International Joint Conference on Pervasive and Ubiquitous Computing (2013)
12. Chuangang, R., Kai, C., Peng, L.: Droidmarking: resilient software water-marking for impeding android application repackaging. In: Proceedings of the 29th ACM/IEEE International Conference on Automated Software Engineering (2014)
13. Kun, Y., Jianwei, Z., Yongke, W.: IntentFuzzer: detecting capability leaks of android applications. In: Proceedings of the 9th ACM Symposium on Information, Computer and Communications Security (2014)

An Audit Log Protection Mechanism
Based on Security Chip

Guan Wang[1,2(✉)], Ziyi Wang[1], Jian Sun[1,2], and Jun Zhou[1]

[1] College of Computer, Beijing University of Technology,
Beijing 100124, China
wangguan@bjut.edu.cn
[2] Beijing Key Laboratory of Trusted Computing, Beijing 100124, China

Abstract. Audit logs can be used to detect the intrusion behavior. So it has become the main target of attack invaders. The existing technologies of logging protection mainly depend on software and have some inherent defects. The actual demand from this, presents an audit logging protection mechanism based on security chip, to provide hardware protection when the log is stored and accessed. Introduction of the security chip makes the audit log to store and access are in the trusted environment, to ensure the confidentiality and integrity of the log.

Keywords: Log security · Security chip · Trust computing

1 Introduction

Audit log has very important role to the safety of the system. It becomes the main target of attack and a subject to various security threats because it can provide evidence of the intrusion of the attacker. The attackers can obtain confidential information through the analysis of logs if the system has been successfully invaded. In order to destroy intrusion traces, the attackers usually forge, tamper or delete the relevant log items and other important data files, and even the entire log file.

Many protection techniques of log have been proposed at home and abroad. Schneier [1] proposed a series of security mechanisms to ensure the security of the log, the main idea is to transfer the log files in the local system to other trusted system to achieve the security of the log, and to use authentication, integrity testing ensure the security of log information transmission. In order to protect the entire log system, intrusion detection is conducted by setting different access policies in the [2], and only the authorized user can access the log and ensure the confidentiality of the log records. At the same time set the log file read and write permissions for access control protection. Chen Xiaofeng [3] proposed a hardware based log protection method. The method is based on the NGSCB platform developed by Microsoft, all the encryption keys are stored in encapsulation, and enhance the security of the key.

Through the analysis of the present research, we can see that there is still a lot of research on the security of audit log. On the one hand, because almost all audit log protection mechanism is based on software implementation, in which encryption and decryption of key is in memory, there exists security risk. On the other hand, because

M. Yung et al. (Eds.): INTRUST 2015, LNCS 9565, pp. 226–233, 2016.
DOI: 10.1007/978-3-319-31550-8_15

of the specific hardware platform is provided by Microsoft, it limits the use of the promotion. Based on the practical requirements, we propose an audit log protection mechanism based on security chip.

This paper first proposes a design method for the structure of log files based on the consideration of confidentiality and integrity. Then the security chip is introduced to ensure the security of the audit log from two aspects: log storage and log access. Finally, analysis of the security and performance is carried out.

2 The Audit Log Protection Mechanism Based on Security Chip

2.1 Objectives of Audit Log Protection

Audit log records the important information of the system running, especially some of the invasion of the trace, only authorized users can access. At the same time as the audit log, only to ensure the integrity of the log record, can play the role of intrusion monitoring.

Considering the above factors, audit log protection mechanism based on security chip should meet the following design objectives.

(a) Confidentiality protection

The log items should be encrypted and stored in the log file when the log is generated. While the log file is accessed, only authorized users can decrypt the log item, to prevent unauthorized users to read audit logs and to master the system's private information. Through the key management and authorization mechanism provided by the security chip TCM, the hardware based log protection is realized.

(b) Integrity protection

The integrity detection mechanism is required to complete the detection of the integrity of the log when accessing the log, so as to determine whether the contents are integrity. Only the integrity of the log is not destroyed is credible. The integrity testing of the log is achieved by using of hash operation module provided by TCM.

2.2   Audit Log Protection Method Based on TCM

To meet the above objectives of the audit log protection mechanism is shown in Fig. 1.

The log protection mainly includes two aspects: the protection of log storage process and the protection of log access process. The two aspects protection ensures that the audit log storage, access is credible, and thus ensures the confidentiality and integrity of the log.

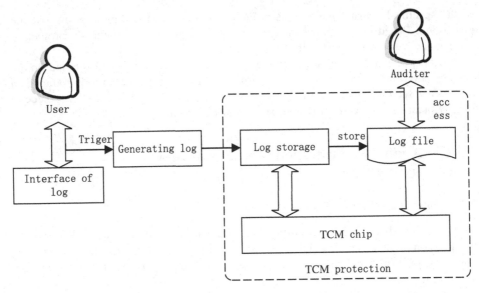

Fig. 1. Audit log protection mechanism based on TCM

3 Design of Log Structure

In general, the log file is stored in the form of clear text without integrity detection mechanism, and can't be checked if it has been tampered with. An attacker is easy to get the system's important information by analyzing the contents of the log file. So if the log files are be protected, it is difficult to understand the log or to forge log data even if the attacker gets the log file data. In order to ensure the confidentiality and integrity of the audit log, the log file structure defined in this paper is shown in Fig. 2.

The log file includes two parts: the encrypted log record and the hash value of log record. When log record is generated, cipher text and hash value of the log record are calculated and are recorded in log file.

In log file, each log record corresponds to a unique serial number, called log record ID, which is generated in a monotonically increasing manner. The log record Id of first log record is 0, followed by 1, and so on. As a result, the log record ID can be used as a judgment basis for the forgery or deletion of a log.

Log record is the core of log files. In order to ensure the confidentiality of the log file, log record should be encrypted with key K. Only audit administrator has the K, therefore only the administrator can decrypt the log records and get the plaintext log.

The hash value of log record is used to verify integrity of log records. In order to guarantee the security of audit logs, the Hash algorithm is provided by TCM, and the hash value is added to the end of the log record.

When the log is read, hash value of each log record is computed and compared with the previous hash value to verify the integrity of the file is destroyed, and to ensure the security of the audit log.

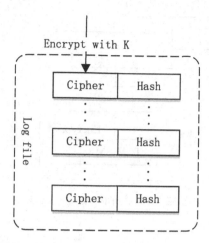

Fig. 2. Structure of log file

4 Log Storage

4.1 Protection of Key

How to store log is the key to guarantee the security of the log. Log encryption is the most effective way to ensure security of log. As a result, the security of the log key determines the reliability of the whole audit log system.

TCM provides cryptographic algorithms, the key uses hierarchical [4] management. The parent key of it must specified at the same time the key is generated, and when the key is stored, read and destroyed, the key must be loaded into the TCM and be encrypted or decrypted with parent key. At the same time, the full use of key authorization mechanism to effectively ensure the security of key storage and use.

In this system, the log key is stored in a permanent storage area provided by TCM. And the key will be assigned a unique identifier UUID. Through UUID, users can easily get the key and perform related operations.

The method of log key security protection is shown in Fig. 3.

4.2 Log Record Encryption

In consideration of the requirements of real-time security log, a fast algorithm of symmetric encryption is selected when the log is encrypted. The log key uses 128 bit SMS4 symmetric key. When system initialize, audit administrator creates log key, whose parent key is the storage main key (SMK), and sets password authorization using log key.

The process for the encryption of the log records and the calculation of the log hash value is shown in Fig. 4.

Firstly, the system obtains the TCM's SMK, and then obtains the key information according to the log key UUID. Since the log key is encrypted by SMK, it is necessary to import the log key into TCM and decrypt the log key by SMK. If the SMK authorization value is correct, TCM uses SMK to decrypt the log key.

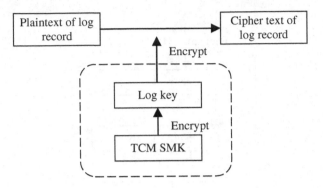

Fig. 3. Method of log key protection

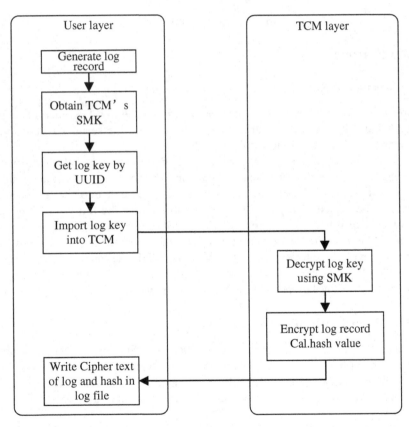

Fig. 4. The process for the encryption and the calculation of the hash value of the log records

Then the log record is encrypted by the log key, and to calculate hash value of log record using Hash algorithm in TCM.

Finally, the cipher text and hash value of log record are written to the log file.

5 Log Record Access

5.1 Log Record Decryption

Since the log record is stored in the form of a cipher text, access log is required to obtain the log key first, then decrypt it. When the log key is created, the system sets up authorization data.

The authorization should be verified before using the log key. At the same time log key is protected by its parent key SMK, it must first import SMK into TCM while decrypting key log. Since SMK is generated in the TCM initialization, and always exists in the chip, it is not the key to be migrated. Therefore, the SMK object is obtained by the authorization of SMK.

When a log key authorized is loaded, the authorization data is imported when the key is created, in addition to log key information and the parent key information. Only the imported authorization data is correct, the access to the log key is obtained. After authentication to SMK and the log key, and obtaining the SMK key, the log the key can be used.

The process for the decryption and the integrity of the log record is shown in Fig. 5.

The audit administrator obtains the key information according to the log key UUID, and obtains the cipher text of log key object. Then the cipher text of log key object is imported into the TCM, and the cipher text of log key is decrypted by the parent key SMK. Read log file, use log key to decrypt log.

5.2 Log Integrity Detection

Integrity detection can detect whether the log information is tampered with or whether the items are inserted or deleted.

Through the log key to decrypt the log records, to calculate the hash value of each log record. To verify the integrity of the log file comparing hash value newly calculated with hash value storing in log record. If two hash value of record are not equal, the log record is tampered.

Each log record corresponds to a unique serial number, and the serial number is recorded in a monotonically increasing manner. If an attacker inserts or deletes a log record, then the number near the insertion or deletion location in the log file is not a monotonic increase.

6 Method Analysis

The protection mechanism uses TCM to realize the log protection based on hardware. The TCM supports the general encryption algorithm, computing power to meet the needs of the system. At the same time, TCM ensures that the system is safe and trusted, so as to ensure that the system log is trusted, and the log encryption key is trusted. TCM uses a hierarchical key to encrypt log record, the log key must be imported into TCM to encrypt and decrypt, thus ensuring that only the authorized users by TCM can use the key.

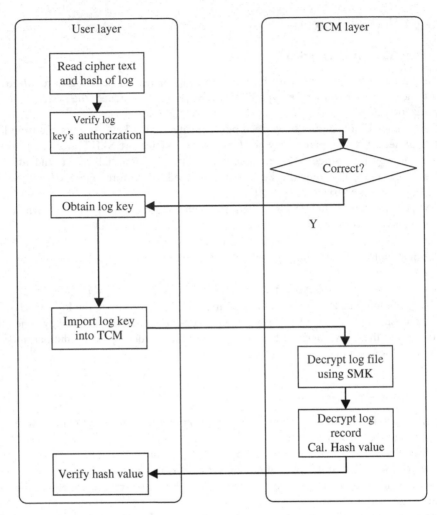

Fig. 5. The process for the decryption and the integrity of the log record

The TCM chip is applied to the system, and the hardware data encryption is realized. Increase encryption and decryption speed, greatly saving the computational expense. The key is also independent of the application program, which can guarantee the security of the key.

7   Conclusions

In this paper, we propose an audit log protection mechanism based on security chip. The mechanism can protect the privacy of the log information, and can find the log records are added or deleted, tampered with. The TCM chip provides log record encryption, decryption and key storage functions. With the help of TCM, ensuring trusted of log encryption process and the secret key storage, so as to improve the strength of log system security.

References

1. Kelsey, J., Callas, J.: syslog-sign Protocol[EB/OL]. http://tools.ietf.org/html. Accessed 1 January 2002
2. Hommes, S., State, R., Engel, T.: A distance-based method to detect anomalous attributes in log files. In: 2012 IEEE Network Operations and Management Symposium (NOMS). IEEE (2012)
3. Xiaofeng, C., Lijuan, Ma., Dengguo, F.: Security audit log based on trusted platform. Comput. Res. Dev. **43**(Suppl), 152–156 (2006)
4. Wujiuxin, A.: Reserch on technique of key management on trusted computing cryptographic support platform. J. Beijing Inf. Sci. Technol. Univ. **24**, 92–96 (2009)

Author Index

Printed in the United States
By Bookmasters